# OXFORD IB COURSE PREPARATION

# PHYSICS

## FOR IB DIPLOMA COURSE PREPARATION

David Homer

OXFORD
UNIVERSITY PRESS

**OXFORD**
UNIVERSITY PRESS

Great Clarendon Street, Oxford, OX2 6DP, United Kingdom

Oxford University Press is a department of the University of Oxford. It furthers the University's objective of excellence in research, scholarship, and education by publishing worldwide. Oxford is a registered trade mark of Oxford University Press in the UK and in certain other countries

British Library Cataloguing in Publication Data
Data available

978-0-19-842359-1

10 9 8 7 6 5 4 3 2 1

Paper used in the production of this book is a natural, recyclable product made from wood grown in sustainable forests. The manufacturing process conforms to the environmental regulations of the country of origin.

Printed in India by Multivista Global Pvt. Ltd
Croydon CR0 4YY

## Acknowledgements

The authors and publisher are grateful to those who have given permission to reproduce the following extracts and adaptations of copyright material:

Ban Ki-moon: Extract from his address to the 66th General Assembly at the United Nations, 21 September 2011. Reproduced by permission of the United Nations.

We have made every effort to trace and contact all copyright holders before publication, but if notified of any errors or omissions, the publisher will be happy to rectify these at the earliest opportunity.

The publisher and the authors would like to thank the following for permission to use their photographs:

**Cover:** NG Images/Alamy Stock Photo

**Artworks:** Thomson

**Photos: p14 (T):** Reproduced by kind permission of the Syndics of Cambridge University Library; PR-ADV-B-00039-00001-000-00046.tif (Adv.b.39.1, p.12); **p14 (B):** H.S. Photos/Alamy Stock Photo; **p23 (L):** Iryna1/Shutterstock; **p23 (C):** Georgios Kollidas/Shutterstock; **p23 (R):** Library of Congress Prints and Photographs Division; LC-USZ62-60242; **p54:** Peter Gould/OUP; **p90:** Bettmann/Getty Images; **p93:** Sybille Yates/Shutterstock; **p99 (R):** Harvard Natural Sciences Lecture Demonstrations; **p99 (C):** Harvard Natural Sciences Lecture Demonstrations; **p99 (L):** Harvard Natural Sciences Lecture Demonstrations; **p112:** Joshua David Treisner/Shutterstock; **p114:** Images of Birmingham Premium/Alamy Stock Photo; **p117:** Artesia Wells/Shutterstock; **p121:** Reprinted with permission from A. Stodolna, M.J.J. Vrakking and co-workers, PhysRevLett.110.213001 (2013); **p127:** Omikron/Science Source/Getty Images; **p148:** Smileus/Shutterstock

# Contents

 Answers to questions in this book can be found at **www.oxfordsecondary.com/9780198423591**

# Introduction to the Diploma Programme

The **Diploma Programme** (DP) is a two-year pre-university course for students in the 16–19 age group. In addition to offering a broad-based education and in-depth understanding of selected subjects, the course has a strong emphasis on developing intercultural competence, open-mindedness, communication skills and the ability to respect diverse points of view.

You may be reading this book during the first few months of the Diploma Programme or working through the book as a preparation for the course. You could be reading it to help you decide whether the Physics course is for you. Whatever your reasons, the book acts as a bridge from your earlier studies to DP Physics, to support your learning as you take on the challenge of the last stage of your school education.

Chapters 1 through to 6 of this book explain the physics that you need to understand at the beginning of a DP Physics course. You may already have met some of this physics, but the book encourages you to begin to look at the concepts that underpin the subject.

Chapter 7 of this book has advice on effective study habits. An appendix lists information you will need throughout the DP Physics course, such as physical constants and equations, maths skills and links to the DP course. The website for this book provides information to help you prepare for tests, examinations and the internal assessment. It also includes answers to questions featured in this book. You can find the URL on the contents page.

## DP course structure

The DP covers six academic areas, including languages and literature, humanities and social sciences, mathematics, natural sciences and creative arts. Within each area, you can choose one or two disciplines that are of particular interest to you and that you intend to study further at the university level. Typically, three subjects are studied at higher level (HL, 240 teaching hours per subject) and the other three at standard level (SL, 150 hours).

In addition to the selected subjects, all DP students must complete three core elements of the course: theory of knowledge, extended essay, and creativity, action, service.

**Theory of knowledge** (approximately 100 teaching hours) is focused on critical thinking and introduces you to the nature, structure and limitations of knowledge. An important goal of theory of knowledge is to establish links between different areas of shared and personal knowledge and make you more aware of how your own perspective might differ from those of others.

The **extended essay** is a structured and formally presented piece of writing of up to 4,000 words based on independent research in one of the approved DP disciplines. It is also possible to write an interdisciplinary extended essay that covers two DP subjects. One purpose of the extended essay activity is to develop the high-level research and writing skills expected at university.

**Creativity, action, service** involves a broad range of activities (typically 3–4 hours per week) that help you discover your own identity, adopt the ethical principles of the IB and become a responsible member of your community. These goals are achieved through participation in arts and creative thinking (creativity), physical exercises (activity) and voluntary work (service).

## DP Physics syllabus
### Basics

The DP Physics course itself is divided into four sections: Core syllabus, Additional Higher Level (AHL) material, and one of four possible Options together with an internal assessment (IA). There is also a group 4 project in which all science students in a school participate.

The Physics course is designed so that a student can study the entire course at standard level with no prior knowledge of physics. At higher level, however, some earlier study of the subject is advisable.

**Physics standard level** = Core + one Option at SL + IA + group 4 project

**Physics higher level** = Core + AHL + one Option at HL + IA + group 4 project

## Physics topics

The *Physics Guide* is a document for teachers that lists the areas that students are to be taught as a series of *Topics*. On page viii there is a table showing the connections between this book and the distribution in the *Physics Guide* of the various areas of physics.

### Core topics

The eight Core topics are taught and examined to the same standard at both levels. They are:

- Measurements and uncertainties
- Mechanics
- Thermal physics
- Waves
- Electricity and magnetism
- Circular motion and gravitation
- Atomic, nuclear and particle physics
- Energy production

Many of these topics and sub-topics are covered in this book.

### AHL topics

There are four additional higher-level topics, which you will only study if you have chosen to study physics at higher level:

- Wave phenomena (an extension of *Waves*)
- Fields (an extension of *Circular motion and gravitation*)
- Electromagnetic induction (an extension of *Electricity and magnetism*)
- Quantum and nuclear physics (an extension of *Atomic nuclear and particle physics*)

### Options

There are four Options, from which you study one. Within each Option some topics correspond to Core areas and one or two additional areas are studied only by HL students. The Options are:

- Relativity
- Engineering physics
- Imaging
- Astrophysics

## Internal assessment (IA)

Physics is an experimental science. The ability to plan and execute an experimental project is part of your assessment, which is where the internal assessment (IA) comes in. The internal assessment may include oral presentations, theoretical investigations and laboratory work. About 10 hours will be devoted to the IA, probably towards the end of the course. Your teacher will support you in carrying out the IA and you will be taught the required skills throughout the course. The IA accounts for 20% of your overall examination marks. There is more advice on the IA on the website of this book (see Contents).

## Group 4 project

Most students who study a group 4 subject undertake a collaborative project within—or possibly beyond—their school. Group 4 students work together, over about 10 hours. The project is not assessed formally. It emphasizes the relationships between sciences and how scientific knowledge affects other areas of knowledge. It can be experimental or theoretical. Be imaginative in your project, and perhaps combine with a different IB World School on another continent to study a project of mutual interest.

### Aims of group 4

There are ten aims addressed by every group 4 subject. Each student should:

- be challenged and stimulated to appreciate science within a global context
- develop scientific knowledge and a set of scientific techniques
- apply and use the knowledge and techniques
- develop experimental and investigative scientific skills
- learn to create, analyse and evaluate scientific information
- learn to communicate effectively using modern communication skills
- realize the value of effective collaboration and communication in science
- have an awareness of the ethical implications of science
- appreciate the possibilities and limitations of science
- understand the relationships between scientific disciplines and between science and other areas of knowledge.

## Key features of the DP Physics course

The following components are incorporated into the DP Physics course:

The **nature of science** (⚛) is the overarching theme in all IB science subjects, including physics. Throughout the course you will encounter many examples, activities and questions that go beyond the studied subject and demonstrate key principles of the scientific approach to exploring the natural world. For example, Isaac Newton—perhaps the most influential physicist of all time—recognized that he was only able to make his ground-breaking discoveries on forces and motion using the work of scientists before him, such as the thought experiments of Galileo Galilei (*1.2 Pushes and pulls*).

Nature of science studies are not limited to the scientific method but cover many other aspects of science, from the uncertainty and limitations of scientific knowledge to the ethical and social implications of scientific research. Raising these issues will help you understand how science and scientists work in the 21st century.

**Theory of knowledge** (🧠) is another common feature of the DP Physics syllabus. In addition to the stand-alone theory of knowledge course taken by all DP students, much of the material in Physics topics can prompt wider discussions about the different ways of knowing used by scientists for interpreting experimental results. For example, the Standard Model was developed to classify nuclear particles (*5.4 The Standard Model*), and proposed the existence of a new particle called a quark. These quarks cannot be observed directly, but using reason and imagination scientists were able to make a case for their existence.

Although theory of knowledge is not formally assessed in the DP Physics course, it facilitates the study of science, just as the study of science supports you in your theory of knowledge course.

**International mindedness** is one of the social aspects of science reflected in the IB mission statement, which emphasizes the importance of intercultural understanding and respect for creating a better and more peaceful world. International mindedness is actively promoted through all DP subjects by encouraging you to embrace diversity and adopt a global outlook. For example, the consequences of ionizing radiation on human health is a global issue (*5.2 Radioactive decay*), and as a result bodies such as the International Atomic Energy Authority exist to promote and encourage nuclear safety throughout the world.

Physics is an experimental science that provides you with numerous opportunities to develop a broad range of practical and theoretical skills.

**Practical skills** (🔧) are required for setting up experiments and collecting data. Typical practical skills are described throughout this book and include using electrical circuits (*2.5 Practical aspects of electrical physics*), using a ripple tank to investigate wave phenomena (*4.1 Waves in theory*), creating models for radioactive decay (*5.3 Half-life*), and convection and conduction experiments (*6.1 Transferring thermal energy*).

**Maths skills** (🖩) are needed for processing experimental data and solving problems. In addition to elementary mathematics, the IB Physics syllabus requires the use of trigonometric functions, radian measure and x-bar notation. Most of the maths skills required for the DP Physics course are outlined in the appendix of this book.

**Approaches to learning** (🎓) are a variety of skills, strategies and attitudes that you will be encouraged to develop throughout the course. The Diploma Programme recognizes five categories of such skills: communication, social, self-management, research and thinking. These skills are discussed in more detail in *7 Tips and advice on successful learning* of this book.

## Assessment overview

In addition to the internal assessment discussed earlier, the **external assessment** is carried out at the end of the DP Physics course. Both HL and SL students are expected to take three papers as part of their external assessment. You will usually take papers 1 and 2 at one sitting with paper 3 a day or two later. The question papers are as follows.*

| Paper | SL | SL timing | HL | HL timing |
|---|---|---|---|---|
| 1 | 30 multiple-choice questions on Core material | 45 minutes; 30 marks; 20% of marks | 40 multiple-choice questions on Core and AHL material | 60 minutes; 40 marks; 20% of marks |
| 2 | Short and extended written answer questions on Core material | 75 minutes; 50 marks; 40% of marks | Short written answer and extended written answer questions on Core and AHL material | 135 minutes; 90 marks; 36% of marks |
| 3 | Section A: Data-based questions and questions based on experimental work<br><br>Section B: Questions on your chosen Option | 60 minutes; 35 marks; 20% of marks | Section A: Data-based questions and questions based on experimental work<br><br>Section B: Questions on your chosen Option | 75 minutes; 45 marks; 24% of marks |

The internal and external assessment marks are combined to give your overall DP Physics grade, from 1 (lowest) to 7 (highest). The final score is calculated by combining grades for each of your six subjects. Theory of knowledge and extended essay components can collectively contribute up to three extra points to the overall Diploma score. Creativity, action, service activities do not bring any points but must be authenticated for the award of the IB Diploma.

## Using this book effectively

Throughout the book you will encounter separate text boxes to alert you to ideas and concepts. Here is an overview of these features and their icons:

| Icon | Feature | Description of feature |
|---|---|---|
| WE | Worked example | A step-by-step explanation of how to approach and solve a physics problem. |
| Q | Question | A physics problem to solve independently. Answers to these questions can be found at www.oxfordsecondary.com/9780198423591 |
| (key) | Key term | Defines an important scientific concept used in physics. It is important to be familiar with these terms to prepare you for the DP Physics course. |
| (DNA) | DP ready – Nature of science | Relates a topic in physics to the overarching principles of the scientific approach to exploring the natural world and the way discoveries are made. |
| (cap) | DP ready – Approaches to learning | Highlights the skills of an effective learner necessary for the DP. |
| (brain) | DP ready – Theory of knowledge | Features ideas or concepts in physics that prompt wider discussions about the different ways of knowing. |
| (link) | Internal link | Provides a reference to somewhere within this book with more information on a topic discussed in the text, given by the section number and the topic name. For example, *6.2 Energy resources* refers to the second section in Chapter 6 of this book and covers different sources of energy. |
| (8) | DP link | Provides a reference to a section of the DP Physics syllabus for further reading on a certain topic. |
| (calc) | Maths skills | Explains an important mathematical skill required for the DP Physics course. |
| (wrench) | Practical skills | Relates the scientific theory to the practical aspects of physics you will encounter on the DP Physics course. |

*Correct at the time of printing

## Linking this book to the DP Physics syllabus

This textbook can be read linearly, but you might find it most useful to dip into specific sections to support different areas of your learning. For example, if you are at the start of your course, you might spend some time reading the section on study skills. Alternatively, if you are learning about electrical theory in class, read through the parts of chapter 2 that explain the conceptual basis of electricity.

The following grid gives a comparison between the chapters of this book and all the IB Diploma Programme Physics course topics. **Bold** indicates that all or part of a sub-topic is covered in this book.

| Topic | Title | Sub-topics | Chapter in this book | Links to Core |
|---|---|---|---|---|
| 1 | Measurements and uncertainties | 1 **Measurements in physics**<br>2 **Uncertainties and errors**<br>3 **Vectors and scalars** | 1 | |
| 2 | Mechanics | 1 **Motion**<br>2 **Forces**<br>3 **Work, energy and power**<br>4 **Momentum and impulse** | 1 | |
| 3 | Thermal physics | 1 **Thermal concepts**<br>2 **Modelling a gas** | 3 | |
| 4 | Waves | 1 **Oscillations**<br>2 **Travelling waves**<br>3 **Wave characteristics**<br>4 **Wave behaviour**<br>5 Standing waves | 4 | |
| 5 | Electricity and magnetism | 1 **Electric fields**<br>2 **Heating effect of electric currents**<br>3 Electric cells<br>4 **Magnetic effects of electric currents** | 2 | |
| 6 | Circular motion and gravitation | 1 **Circular motion**<br>2 Newton's law of gravitation | 1 | |
| 7 | Atomic, nuclear and particle physics | 1 **Discrete energy and radioactivity**<br>2 **Nuclear reactions**<br>3 **The structure of matter** | 5 | |
| 8 | Energy production | 1 **Energy sources**<br>2 **Thermal energy transfer** | 6 | |
| 9 (AHL) | Wave phenomena | 1 Simple harmonic motion<br>2 Single-slit diffraction<br>3 Interference<br>4 Resolution<br>5 Doppler effect | not covered | Topic 4 |
| 10 (AHL) | Fields | 1 Describing fields<br>2 Fields at work | not covered | Topics 6.1 and 6.2 |
| 11 (AHL) | Electromagnetic induction | 1 **Electromagnetic induction**<br>2 **Power generation and transmission**<br>3 Capacitance | 2, 6 | Topic 5.4 |
| 12 (AHL) | Quantum and nuclear physics | 1 The interaction of matter with radiation<br>2 Nuclear physics | not covered | Topic 7 |
| **Option** | | **SL and HL topics** | | **AHL topics (HL only)** |
| A | Relativity | 1 Beginnings of relativity<br>2 Lorentz transformations<br>3 Spacetime diagrams | | 4 Relativistic mechanics<br>5 General relativity |
| B | Engineering Physics | 1 Rigid bodies and rotational dynamics<br>2 Thermodynamics | | 3 Fluids and fluid dynamics<br>4 Forced vibrations and resonance |
| C | Imaging | 1 Introduction to imaging<br>2 Imaging instrumentation<br>3 **Fibre optics** | 3 | 4 **Medical imaging** |
| D | Astrophysics | 1 Stellar quantities<br>2 Stellar characteristics and evolution<br>3 Cosmology | | 4 Stellar processes<br>5 Further cosmology |

# 1 Motion and force

> " The main purpose of science is simplicity and as we understand more things, everything is becoming simpler. "
>
> **Edward Teller, *Conversations on the Dark Secrets of Physics* (1991)**

## Chapter context

This chapter deals with how things move and introduces you to some important related concepts that extend throughout physics—in particular, the **conservation of momentum**.

## Learning objectives

In this chapter you will learn about:

→ **distance** and **displacement**, **speed** and **velocity**, and **acceleration**

→ **displacement–time** and **velocity–time graphs**

→ the **kinematic (*suvat*) equations**

→ forces and **Newton's three laws of motion**

→ the effects of **friction**

→ **work, energy** and **power**

→ conservation laws

→ **momentum** and **impulse**.

 **Key terms introduced**

→ displacement
→ velocity
→ acceleration
→ force
→ Newton's laws of motion
→ work done
→ power
→ momentum
→ impulse

## 1.1 Faster and faster

There are two aspects of motion to study—how it is defined and measured, and how it can be changed. This first section looks at the basic definitions and how they are linked.

**DP link**

In the IB Physics Diploma Programme you will learn about units when you study **1.1 Measurements in physics**.

You will learn about distance and displacement in **2.1 Motion**.

---

**DP ready** **Nature of science**

### Mass, length and time

All cultures measure the quantities used in everyday life, but they use many different units; for example, mass is measured in grams, ounces (to measure gold) and maunds (an Indian unit). Science uses a single agreed system of units that was established in 1960, though its development began well before that. It is called the SI (*Système Internationale d'unités*). Seven base units are defined as **fundamental** units. All other units are derived from these and are **secondary** units.

The fundamental SI units used in this book, together with their abbreviation and quantity, are:

- mass: the kilogram (kg)
- length: the metre (m)
- time: the second (s)
- electric current: the ampère (A)
- amount of substance: the mole (mol)
- temperature: the kelvin (K)

A fundamental unit not used in the Physics Diploma Programme is the candela (cd), the unit of luminous intensity.

## Distance and displacement

Imagine running round an athletic track. Figure 1 shows a map of the track. The total distance around the track on the inside lane is 400 m. When you have run halfway round, how far have you travelled?

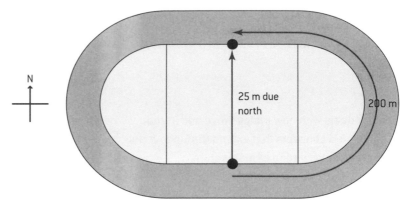

**Figure 1.** Distance and displacement on a race track

One answer is 200 m—half the track length. This is the *distance* you have run. It is a measure of how much ground you have covered irrespective of direction. It is a *scalar* quantity.

But another way to look at it is that you are only 25 m away from your starting point and due north of where you began. This is your *displacement*, which is a *vector* quantity, and always requires a magnitude (the number part) and a direction (the start-to-finish information).

Continue running back to the starting point. Your distance travelled is now 400 m but your displacement has become zero.

### Key term

**Distance** is the length of a path travelled between two points. It is a scalar quantity.

**Displacement** is the difference (in magnitude and direction) between an initial and final position. It is a vector quantity.

The units of distance and displacement are the metre (m).

A **scalar** quantity has only magnitude; a **vector** quantity has both magnitude and direction.

### Internal link

Vectors and scalars occur in many topics, notably in the kinetic theory of gases (**3.2 Gas laws**).

### DP link

In the IB Physics Diploma Programme, ideas about motion are studied in **2.1 Motion**, and vectors are studied in **1.3 Vectors and scalars**.

### Question

1 a) Calculate for your journey from home to school
   i) your displacement (including direction)
   ii) your distance travelled.
  b) Identify how your answers to **(a)** change for your journey going from school to home.

### Maths skills: Scalars and vectors

You need to know how to manipulate scalars and vectors. Here are the ground rules:

**Adding and subtracting:** Scalars are numbers, and are added and subtracted like ordinary numbers.

In adding or subtracting vectors you must take account of the direction as well as the size. The best way to see this is to begin with a scale drawing. Imagine that a boy cycles 3 km due north along a straight road and then 4 km along another road that goes due east (figure 2).

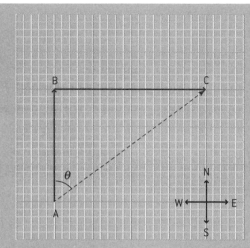

**Figure 2**

**Table 1.** Examples of scalars and vectors

| Examples of scalar quantities | Examples of vector quantities |
|---|---|
| mass | weight |
| speed | velocity |
| time | acceleration |
| energy | force |
| power | magnetic field strength |
| temperature | electric current |

From figure 2, the total distance travelled is $(3 + 4) = 7$ km. What is the displacement? It is measured from the beginning of the journey, A, direct to the end of the journey, C. There are two ways to work this out.

One way is to use trigonometry. Compute the distance using Pythagoras's theorem: $\sqrt{3^2 + 4^2} = 5$ km

and compute the angle using $\theta = \tan^{-1}\left(\dfrac{4}{3}\right) = 53°$.

So the displacement is 5.0 km in a direction N 53°E.

The other way is by scale drawing. Draw the first vector upwards (north), 3 cm long (using the scale 1 km ≡ 1 cm). At the top end of this vector (which shows where the boy was after the first leg of the journey) draw a second line. This should be $4 \times 1$ cm, that is, 4 cm long, and should go to the right. Use a protractor (or squared paper) to ensure that the angle between the vectors is 90°. The displacement is the vector (called the resultant vector) that stretches from the start of the first vector to the end of the second vector. Measure it and it should be 5 cm long; use a protractor to check that the angle between the first vector and this resultant is 53°.

To subtract vectors by scale drawing, treat the vector being subtracted as though it had the opposite direction to its actual direction. Then add this new (negative) vector to the other.

The idea of adding a scalar to a vector in physics has no meaning. It is like adding an energy in joules to a velocity in metres per second.

**Multiplying and dividing:** Again, scalars are multiplied and divided just like ordinary numbers.

It is possible to multiply a vector by a scalar. The direction does not change, and the magnitude of the vector is multiplied by the scalar. A velocity of 10 m s$^{-1}$ in a direction due east that is multiplied by 5 becomes 50 m s$^{-1}$ still in the direction due east.

There are two ways to multiply vectors together (called "dot" and "cross" products); you may meet them in the IB Mathematics Diploma Programme, but they will not be required in Physics.

 **Internal link**

There are some notes on trigonometry at the end of **1.2 Pushes and pulls**.

## Key term

**average speed**

$$= \frac{\text{total distance travelled}}{\text{total time taken}}$$

**average velocity**

$$= \frac{\text{change in displacement}}{\text{time taken}}$$

The units of speed and velocity magnitude are metres per second (m s⁻¹; it is not usual to write m/s even though you might have done so in earlier study). Speed is a scalar quantity; velocity is a vector and always needs both magnitude and direction.

## DP link

You will learn about significant figures and their treatment when you study **1.2 Uncertainties and errors**.

**Table 2.**

| Time (s) | Distance from start (m) |
|----------|--------------------------|
| 0 | 0 |
| 1 | 2 |
| 3 | 16 |
| 5 | 44 |
| 7 | 86 |
| 9 | 142 |
| 11 | 205 |
| 13 | 275 |
| 15 | 345 |
| 17 | 415 |

## Average speed and velocity

We often need to know not just the length of our journey but also how quickly we travelled. To do this we define two quantities that mirror the distance and displacement quantities: these are *speed* and *velocity*. For the calculation of speed or velocity we often use the time average. Time here is the travel time between measuring the first position and the second position.

### Worked example: Speed and velocity

1. Look again at the running track shown in figure 1. If it takes you 40 s to run halfway round, you cover the distance with an average speed of $\frac{200\text{ m}}{40\text{ s}} = 5.0\text{ m s}^{-1}$. But the average velocity is $\frac{25\text{ m}}{40\text{ s}} = 0.63\text{ m s}^{-1}$ due north.

### Maths skills: Significant figures, decimal places and standard form

The running track problem has its answers expressed to two significant figures (2 sf) because this was the smallest number of sf expressed in the data. Writing "200 m" implies that we know the value to the nearest metre, that is, 200 ± 1 m; this is 3 sf, whereas "40 s" implies that we know the time to the nearest second (2 sf).

Never quote an answer to better than the smallest number of sf in the data. And be careful with rounding when you adjust the final answer.

Decimal places (dp) are often confused with significant figures: 123.45 is a value quoted to 5 significant figures and 2 decimal places.

A good way to avoid being tripped up by sf and dp is to use standard form: $1.2345 \times 10^3$. When you deal with small or very large numbers such as the mass of a proton ($1.67 \times 10^{-23}$ kg to 3 sf), standard form is crucial.

### Question

2 A teacher walks 5 m north, 2 m east, 5 m south and 2 m west. The whole journey takes 42 s. Calculate the teacher's
   **a)** average speed **b)** average velocity.

3 Give two examples of a **vector** quantity and two examples of a **scalar** quantity.

## Instantaneous speed and velocity: distance–time graphs

Car drivers realize that an important fact about a car journey is not necessarily the average speed, but the speed that a roadside camera records! This is known as the **instantaneous speed**, the speed at one moment in time. For a car, it is the speed indicated by the speedometer.

Graphs make it much easier to visualize speeds compared to data tables. To demonstrate, consider the data for the distance travelled by a car in a straight line during the first few seconds of a journey in table 2. These data could be laboriously transformed into a set of average speeds by working out the distance travelled between successive pairs and dividing by the time between them, but plotting the graph from the distance–time data shows details of the motion straight away.

Figure 3 shows the data plotted as a graph of distance (y-axis) against time (x-axis) with the best-fit curve drawn.

The car moves slowly at the start, so the gradient of the graph is small. As time goes on the speed increases (the graph is steeper) until it becomes constant (a straight line beyond 10 s).

The instantaneous speed at a particular time can be determined from a distance–time graph by finding the gradient of the line at that time.

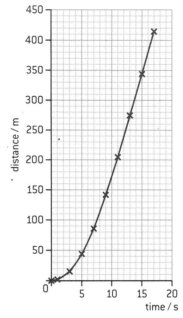

**Figure 3.** Distance–time data from table 2

## Maths skills: Calculating a gradient

The technique below applies to finding the gradient of any graph at a point, whether a straight line or a curve. This example is a distance–time graph, and we require the instantaneous speed at a time of 7.0 s.

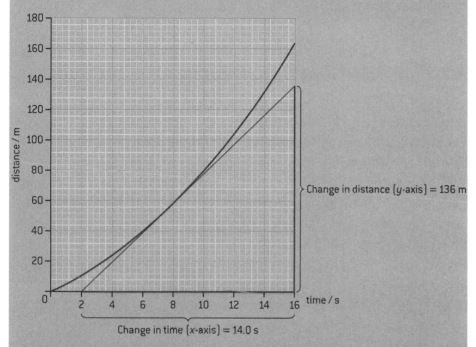

Change in distance (y-axis) = 136 m

Change in time (x-axis) = 14.0 s

If the graph is a curve, draw a tangent to the line at 7.0 s (if the graph is a straight line then this step is not needed). The tangent line should be as long as possible.

Read off the intercepts on the axes and work out the gradient from:

$$\frac{\text{change in the } y\text{-direction}}{\text{change in the } x\text{-direction}}.$$

The values for this example are on the graph.

Treating this as an equation, you will see that the

$$\text{gradient} = \frac{\text{change in distance / m}}{\text{change in time / s}} = \text{speed, measured in m s}^{-1}$$

$$= 9.7, \text{ measured in m s}^{-1}.$$

Always quote the quantity and the unit in the final answer for a gradient (so 9.7 m s$^{-1}$).

**4** The distance–time graphs below show the motion of three objects, A, B and C.

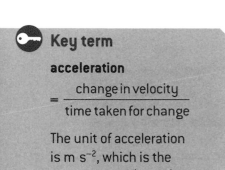

A
distance / m

B
distance / m

C
distance / m

Describe the motion of the objects as fully as you can.

**5** A car travels at a steady speed of 22 m s⁻¹ for 20 minutes. Then the car goes a further distance of 35 km for 15 minutes.

Calculate:

a) the distance travelled in the first 20 minutes

b) the average speed for the 35 minute journey.

## Key term

**acceleration**

$$= \frac{\text{change in velocity}}{\text{time taken for change}}$$

The unit of acceleration is m s⁻², which is the equivalent of $\frac{(m\,s^{-1})}{s}$.

**Table 3.**

| Time (s) | Speed (m s⁻¹) |
|----------|---------------|
| 0.0 | 0.0 |
| 1.0 | 1.5 |
| 2.0 | 3.0 |
| 3.0 | 4.5 |
| 4.0 | 6.0 |

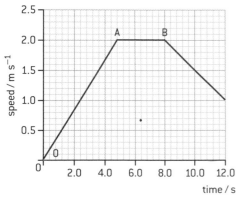

**Figure 4.** Speed–time graph

## Acceleration: speed–time graphs

The final quantity used to describe motion is *acceleration*. This is a measure of the rate at which velocity changes. The word "rate" is another way to say "change in [quantity] per unit time".

Table 3 gives data of the speed every second for a car that moves from rest. After one second the car goes from 0 to 1.5 m s⁻¹ so the change in speed in the first second is 1.5 m s⁻¹.

In the time from 1 s to 2 s the change in speed is again 1.5 m s⁻¹ (= 3.0 − 1.5).

In the third second (2 s to 3 s), the speed change is (4.5 − 3.0), still 1.5 m s⁻¹.

So for this journey, in every second of the motion, the speed increases by 1.5 m s⁻¹ per second. The change in speed is 1.5 m s⁻¹ per second; this is an acceleration of 1.5 (m s⁻¹) s⁻¹, written as 1.5 m s⁻².

Acceleration is a vector quantity (with direction), though you will not always know a direction. If in doubt as to what is needed, always assume it is a vector and quote a direction if possible.

Again, there is a distinction between average acceleration (the change in speed each second over a definite time interval) and the instantaneous acceleration (the change in speed each second at one instant in time). And again, a graph shows these distinctions (figure 4).

First, look at the overall shape of the graph and see what it shows: the object starts at rest (meaning it has zero speed at zero time). Then the speed increases steadily for the first 4.8 s. The gradient of this straight line (region OA) gives the acceleration. From 4.8 s to 8.0 s (region AB) the speed does not change; the gradient of the graph, and therefore the acceleration, is zero. From 8.0 s to 12.0 s the speed is decreasing, so the acceleration now has a negative value.

One term often used to describe a decrease in speed is "deceleration". Take some care with this: it is better to call the quantity "acceleration" and then to use a minus sign to make it clear that the gradient of the velocity–time graph (and therefore the acceleration) is negative.

## Speed–time graphs and distance travelled

There is more information to be gained from a speed–time graph such as figure 4.

As discussed before, the quantities speed and time give acceleration from $\frac{\text{speed}}{\text{time}}$. But notice that the definition for speed can be rearranged to give distance = speed × time. The quantity (speed × time) represents the area under a speed–time graph. We can work out the distance travelled for part or all of a journey by calculating the area under the line for the speed–time graph. The units for speed × time are (m s⁻¹) × (s). The units of seconds cancel, leaving only metres, as you should expect.

### Maths skills: Estimating the area under a speed–time graph

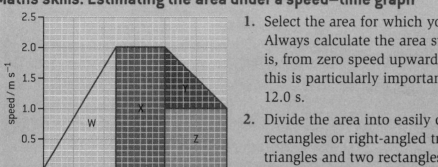

**Figure 5.** Calculating area by the use of regular shapes

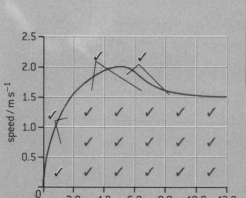

**Figure 6.** Counting squares to estimate area

1. Select the area for which you need to know the distance. Always calculate the area starting from the time axis (that is, from zero speed upwards). In the example in figure 5, this is particularly important for times between 8.0 s and 12.0 s.

2. Divide the area into easily calculated regions, either rectangles or right-angled triangles. In figure 5, two triangles and two rectangles do the job. When there is a curved line you may have to estimate the area (figure 6).

3. Either count the squares in the grid (best for curves) or calculate the area (best for lines). Remember that the area of a triangle is $\frac{1}{2} \times$ base × height whereas a rectangle is base × height.

4. Add together all the areas to get the total distance.

**Table 4.** Area calculation for the speed–time graph in figure 5

| Area | Calculation | Distance / m |
|---|---|---|
| W | $\frac{1}{2} \times (4.8 - 0) \times 2.0$ | 4.8 |
| X | $(8.0 - 4.8) \times 2.0$ | 6.4 |
| Y | $\frac{1}{2} \times (12.0 - 8.0) \times (2.0 - 1.0)$ | 2.0 |
| Z | $(12.0 - 8.0) \times 1.0$ | 4.0 |
| **Total** | | **17.2** |

In the example the total distance travelled is 17.2 m—probably best expressed as 17 m to 2 sf.

If you cannot divide the area into regular shapes, then count the number of squares, as shown in figure 6.

There is a tick in every complete large square, and some ticks where incomplete squares are roughly equivalent to one large square, making about 19 squares altogether. An estimate to the nearest square is as good as you will be able to manage. Each square is 0.5 m s⁻¹ by 2.0 s in area, in other words 1.0 m. So, 19 m underneath the graph in total.

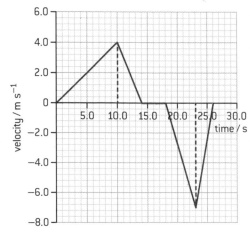

**Figure 7.** Velocity–time graph

## Velocity–time graphs and displacement

A velocity–time graph can give even more information, this time including direction. Figure 7 shows a graph for a journey along a **straight line** (we need to know this, otherwise we cannot make some of the later deductions). As usual the gradients of the graph give the accelerations (also in the direction of motion). This time, however, the line goes below the *x*-axis. When it does so, the velocity is negative. This means that the object is now travelling back towards the starting point. The area is also negative and represents displacement back towards the starting position.

### Worked example: Analysing velocity–time graphs **WE**

2. Analyse as much as you can of the motion for figure 7.

| Time / s | Analysis |
|---|---|
| 0–10 | Accelerating; acceleration is 0.40 m s$^{-2}$; displacement is 20 m in +ve (positive) direction |
| 10–14 | Slowing down to zero; acceleration is −1.0 m s$^{-2}$; displacement is 8 m in +ve direction |
| 14–18 | Stationary; no change in displacement |
| 18–23 | Accelerating but towards starting point; −1.4 m s$^{-2}$; displacement is −17.5 m |
| 23–26 | Slowing down so accelerating in +ve direction; +2.3 m s$^{-2}$; displacement is −10.5 m |
| 0–26 | Displacement is +20 + 8 − 17.5 − 10.5 = 0 m so object arrives back at starting point |

### Question **Q**

6 The graph shows the variation of speed with time for a car.

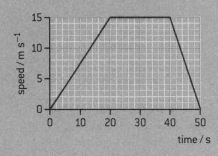

a) State the maximum speed of the car.

b) Calculate the acceleration for

    i) the first 20 s of the motion

    ii) the last 10 s of the motion.

c) Determine the distance travelled in the first 40 s.

d) Determine the average speed for the whole journey.

7 A series of speed–time graphs are shown for four different journeys, A, B, C and D.

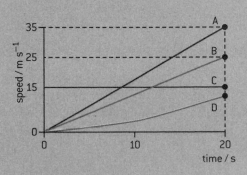

a) Compare the journeys of A and B in as much detail as you can.

b) Describe journey C.

c) Describe journey D.

## Kinematic equations

Speed–time graphs are a good way to visualize motion, and to estimate acceleration and distance travelled. However, sometimes there is a better method for calculating either the speed or acceleration or distance travelled: the **kinematic equations**. These equations are sometimes called the *suvat* equations from the symbols used:

| | |
|---|---|
| *s* | **distance** travelled |
| *u* | initial **speed** |
| *a* | acceleration |
| *t* | **time** taken |

*v*   final **speed**

To use these equations we assume that the **acceleration is constant** and does not change throughout the motion. The acceleration is said to be **uniform** when this is true.

> **DP link**
>
> You will learn about the kinematic (*suvat*) equations in **2.1 Motion**.

---

**DP ready**  **Nature of science**

The four kinematic equations make assumptions about the systems they describe. The most important is that the acceleration is constant. When this is not true, the equations are not valid (and you may be penalized in an examination for using them). An example is a skier moving down a hill with a varying slope. The acceleration down the slope will not be constant so the equations do not apply.

Another assumption is that we are dealing with point objects. We do not consider the mass or distribution of mass of the objects.

These equations apply to translation only, not rotation (though they can be extended to rotation, as you will learn if you study Option B of the IB Diploma Physics Programme).

---

The speed–time graph below (figure 8) shows the change in speed of an object over time *t*. Compare the starting and finishing speeds with the list of symbols above. The graph is a straight line and, of course, this tells us that the acceleration is constant. The derivations for the four equations, related to the graph, are shown below.

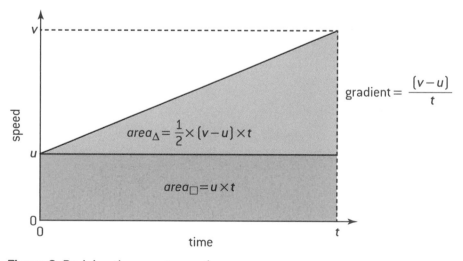

**Figure 8.** Deriving the *suvat* equations

## Equation 1

The acceleration is the gradient of the graph, so

$$\text{acceleration, } a = \frac{\text{change in speed}}{\text{time for speed change}} = \frac{(v-u)}{(t-0)}$$

$$a = \frac{v-u}{t} \text{ and therefore } v = u + at.$$

**Equation 2**

The distance travelled (area under the graph) can be evaluated as two areas: $\text{area}_\triangle$ and $\text{area}_\square$.

The total area, $s$ (distance travelled), is

$$\text{area}_\triangle + \text{area}_\square = \frac{1}{2} \times (v - u) \times t + u \times t = \frac{1}{2} \times at \times t + u \times t$$

(this also uses $v = u + at$ rearranged as $v - u = at$)

which becomes $s = ut + \frac{1}{2}at^2$.

**Equation 3**

Combine equations 1 and 2 and eliminate $t$ to give $v^2 = u^2 + 2as$.

**Equation 4**

Eliminate $a$ from the others to give $s = \left(\dfrac{v + u}{2}\right)t$.

---

**Worked example: Using the kinematic (*suvat*) equations**

3. A car accelerates uniformly along a straight road, taking 13 s to change its speed from 8.0 m s⁻¹ to 34 m s⁻¹. Calculate

   **a)** the acceleration of the car

   **b)** the distance travelled by the car in the 13 s time period.

*Solution*

    **a)** Begin by writing down what you know and what is required.

        $t = 13$ s

        $u = 8.0$ m s⁻¹

        $v = 34$ m s⁻¹

        $a = ?$

        so $a = \dfrac{v - u}{t} = \dfrac{(34 - 8)}{13} = 2.0$ m s⁻²

    **b)** $s = ut + \dfrac{1}{2}at^2 = 8 \times 13 + \dfrac{1}{2} \times 2 \times 13^2 = 104 + 169 = 273$ m

        As all the data are to 2 sf, the distance travelled is best given as 270 m.

4. An aircraft lands on a runway, taking 920 m to stop from a landing speed of 45 m s⁻¹. Calculate

   **a)** the time to stop

   **b)** the average deceleration.

*Solution*

    **a)** $s = 920$ m

        $u = 45$ m s⁻¹

        $v = 0$

        $t = ?$

        Equation 4 can be rearranged as $t = \dfrac{2s}{v + u} = \dfrac{2 \times 920}{0 + 45} = 40.9$ s or 41 s to 2 sf.

    **b)** One route is to use $v^2 = u^2 + 2as$

        So $0^2 = 45^2 + 2 \times a \times 920$         (it is important to link the values to the symbols: $45^2 = 0^2 + 2 \times a \times 920$ is wrong)

        $a = -\dfrac{2025}{2 \times 920} = -1.1$ m s⁻² to 2 sf         (notice the minus sign; it tells us that the aircraft is slowing down, so this is a deceleration)

## Question

8   A motorcyclist accelerates uniformly from rest to a speed of 45 m s$^{-1}$ in 12 s. Then she brakes with a uniform deceleration to stop in a distance of 85 m.

   **a)** Calculate, for the first 12 s of the journey,

      **i)**   the acceleration

      **ii)**  the distance travelled.

   **b)** Calculate, for the second part of the journey,

      **i)**   the deceleration

      **ii)**  the time taken to stop.

   **c)** Sketch a graph to show the variation of speed with time for this journey.

   **d)** Use the graph to calculate the average speed for the whole journey.

## Acceleration due to gravity

When an object falls from rest close to the Earth's surface, it accelerates downwards. The magnitude of this acceleration due to gravity, given the symbol $g$, can be measured by dropping a small ball from rest below an ultrasound sensor connected to a data logger.

**Table 5.** Averaged results for the experiment in figure 9

| Time (s) | Speed (m s$^{-1}$) | Time (s) | Speed (m s$^{-1}$) |
|---|---|---|---|
| 0 | 0 | 0.35 | 3.64 |
| 0.05 | 0.45 | 0.40 | 3.64 |
| 0.10 | 1.04 | 0.45 | 4.10 |
| 0.15 | 1.36 | 0.50 | 4.55 |
| 0.20 | 1.95 | 0.55 | 5.36 |
| 0.25 | 2.60 | 0.60 | 0.00 |
| 0.30 | 3.12 | 0.65 | 0.00 |

Data loggers can usually be programmed to produce either a distance–time graph or a speed–time graph; the latter gives more information.

Table 5 gives the averaged speed–time results for three runs of this experiment. The results for 0.60 s and beyond show that the ball must have stopped moving somewhere between 0.55 s and 0.60 s. It probably hit the bench.

To find the value of $g$:

1. Begin by drawing the graph and then constructing the best-fit line (there is advice in the Maths skills section on page 12). Notice that there are some random errors in the measurements.

2. Measure the gradient and use it to calculate $g$. Compare your answer with the accepted value.

3. There is more you can find out from this graph. Think about the other quantity that a speed–time graph can give. What will it tell you in this experiment?

### DP link

You will learn about the acceleration due to gravity and its determination in **2.1 Motion** and in **6.2 Newton's law of gravitation**.

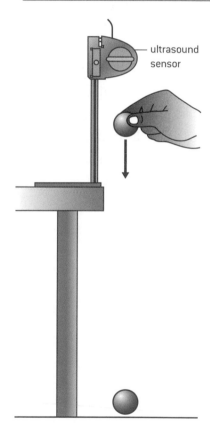

ultrasound sensor

**Figure 9.** Measuring the acceleration due to gravity, $g$

## Maths skills

### Plotting graphs

- Use sensible scales for your axes: 1:1, 1:2, 1:5 are good; 1:3, 1:6, 1:7 and 1:9 are hard to use.
- A graph should occupy *at least* half the grid on the graph paper.
- To achieve the point above, consider using a false origin (one that is not (0,0)).
- Mark your data points consistently and clearly, use ×, +, ⊙.
- All marks on the graph (plots or lines) should be drawn with a sharp pencil.
- Label axes correctly with the quantity / power of ten and unit, for example distance / $10^3$ m.

### Drawing a best-fit line

- Draw straight lines with a transparent ruler (so you can see all the points at once).
- Draw curves free-hand, in one movement that you have practised several times without putting the pencil to paper. Turn the paper before you start so that your hand is on the inside of the curve.
- Get a balance of points on each side of the line (whether straight or curved). Make the total distance from points to the line as small as possible.
- If there are error bars on the data, draw the line through all the error bars if possible.
- Don't force the line through the origin unless you are sure this is the correct physics for the situation.

## Question

Q

9   A cyclist accelerates uniformly from rest to a speed of 9.0 m s$^{-1}$ in a time of 45 s. Then he immediately applies the brakes and stops with uniform acceleration. The braking distance is 27 m.

   a) Calculate, for the first 30 s,

      i)   the acceleration                     ii)  the distance travelled.

   b) Calculate, for the braking,

      i)   the acceleration                     ii)  the time taken to come to rest.

   c) Determine the average speed for the whole journey.

10  Figure 10 shows the speed–time graph for a sprinter in a race.

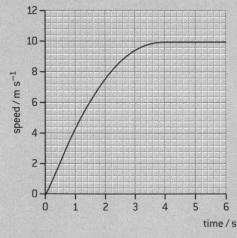

Determine

a) the acceleration of the sprinter at the start of the race

b) the total distance travelled in 6.0 s

c) the average speed of the sprinter over the first 4.0 s.

**Figure 10.** Speed–time graph for a sprinter

## 1.2 Pushes and pulls

You may have been taught that a *force* is a push or a pull that acts on something due to another object. In this section you will look at forces: what they are and what they do.

### Balanced forces

Imagine a ball resting on a table on the Earth's surface (figure 11). The ball is not moving relative to the table or the Earth even though the gravitational pull of the Earth and other forces are acting on it. This is because all the forces are **balanced**. We say that the ball is in **equilibrium**.

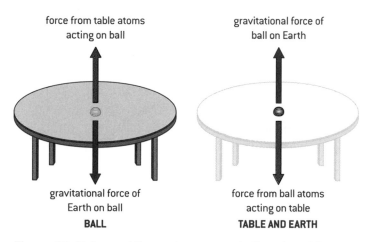

**Figure 11.** Balanced forces between a ball and a table

However, a careful examination of the forces shows that the situation is more complex than this. As well as the gravitational effects, the surface of the table and the ball are deformed slightly by the gravitational forces that are acting. The diagram shows these four forces (the ball and the table are separated for clarity):

- the weight of the ball (the Earth's downward gravitational pull on it)
- the gravitational force of the ball on the Earth upwards (this has a tiny effect as the Earth is so large, but it exists)
- the spring force of the table upwards on the ball as the surface tries to return to being flat
- the spring force of the ball downwards on the table as the ball tries to return to being spherical.

The gravitational forces and the spring forces are balanced, so the net force (all the forces added together) is zero.

A crucial point to recognize here is that the directions of the forces are discussed as well as their magnitude. Forces are vectors and have both magnitude and direction.

 **DP link**

You will learn about forces and Newton's laws of motion when you study **2.2 Forces**.

 **Key term**

When a **force** acts on an object, the object moves if it is free to do so. More precisely, the force causes an acceleration. Unless, that is, some other force prevents the motion.

 **Internal link**

Balanced forces are discussed further in the context of Newton's third law of motion later in this section, and frictional forces are covered in **Friction effects** at the end of this section.

Some more examples of cases in which forces are balanced include:

- an ice cube floating at rest in a glass of water
- an aircraft moving at a constant velocity, where the thrust and the air resistance are exactly in balance
- a child pulling a sled at a constant velocity along snow (the tension in the rope to the sled is equal to the friction at the snow surface).

## Newton's first law of motion

If you live in a part of the world that is very cold in winter, you will be very familiar with the sled example above. When a sled is pushed on a horizontal surface it can travel for a long distance before stopping; the frictional force of the ice on the runners is small. What would happen if there were no friction at all? The answer is that the sled would continue to move at a constant velocity. This is the basis of *Newton's first law of motion*. A net force must act before an object's velocity can change (that velocity could be zero if the object is initially stationary). The use of velocity rather than speed here is crucial because, as you will see when you study circular motion, the direction of the force vector relates to the direction of the vector change in the velocity.

**Figure 12.** A page from Newton's *Principia* in which he discusses the three laws of motion (and much more)

---

**DP ready** | **Nature of science**

### Galileo and Newton

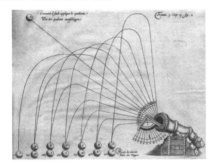

**Figure 13.** Not how cannonballs really travel

Newton was not the first to recognize the relationship between force and change in velocity. Galileo and others were beginning to come to this conclusion in Europe during the 16th century. Before then people thought that force had constantly to be supplied to enable an object to keep moving. This picture, drawn by Diego Ufano, a Spanish military engineer who died in 1613, shows how people once thought cannon balls moved in the air, running out of "force" just before they fall vertically to the ground (figure 13).

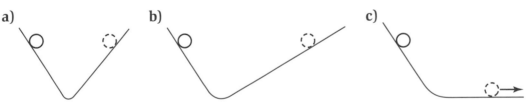

**Figure 14.** Galileo's thought experiment

Galileo performed a "thought experiment" to help himself see what was happening. He imagined a ball, released from rest, on a V-shaped ramp (figure 14). The ball reaches the same height on the right-hand side of the ramp in a) and b) if no friction acts. What happens if the right-hand side of the V is made horizontal? Galileo realized that the ball would roll on for ever—no force, no change in velocity.

Newton recognized the importance of the work of earlier scientists to his own thinking. He said: "If I have seen further, it is by standing on the shoulders of giants"; he was probably using the idea of 12th-century philosopher Bernard of Chartres, who realized that truth almost always builds on previous discoveries.

## Newton's second law of motion

Force, mass and acceleration are related by:

force = mass × acceleration    $F = ma$

- Notice that $m$ is a scalar quantity and $a$ is a vector; this is permitted, because the mass quantity simply multiplies the vector and does not change its direction.

- A consequence of this is that the direction of $a$ and the direction of $F$ are the same.

- Only one $F$ is referred to in the equation. This is the **resultant force** or total force if there are two or more forces. You met the idea of adding vectors in *1.1 Faster and faster* in the "Scalars and vectors" section. You can use the drawing or calculation method for forces too.

---

### Worked example: Newton's second law of motion

5. A car of mass 900 kg accelerates from rest to 15 m s⁻¹ in 50 s. Calculate the resultant force acting on the car.

**Solution**

Using $v = u + at$, $a = \dfrac{15 - 0}{50} = 0.30$ m s⁻²

So $F = 900 \times 0.30 = 270$ N

---

## Mass and weight

A gravitational force acts on any object in the gravitational field of another. This is usually only obvious to us when we are considering how the pull of gravity—the gravitational force due to the Earth—acts on us or any other object on the Earth: in other words, *weight*. In fact, all masses exert a gravitational force on each other. The larger the mass, the larger the force. The forces are very small unless we are dealing with something the size of a planet or a moon.

For a mass of 1 kg, $F = $ mass $\times g = 1 \times 9.81 = 9.81$ N (roughly 10 N).

---

**DP ready**   **Theory of knowledge**

**Is mass constant?**

Einstein, in his special theory of relativity, said mass was not a constant—and experimental evidence now backs up this theory. Einstein showed that observed mass increases when speed increases. If you study Option A of the IB Physics Diploma Programme, you will look at this phenomenon in detail. However, for most of the course, you can assume that mass is constant.

---

So, if weight is gravitational pull, what is mass? This is the amount of substance in an object. However, although this definition is correct, it is not very helpful. In fact, it is difficult to pin down the concept of mass exactly other than to say it is the quantity that responds to force by accelerating. You can regard mass as a constant that depends on the number of atoms in an object, whereas weight can vary over the Earth's surface (because the value of $g$ varies over the surface) and if someone makes measurements on the Moon or a nearby planet.

### Internal link

We will see another way to write the law in **1.4 Momentum and impulse**.

### Key term

**Newton's second law of motion** states that $F = ma$ where $F$ is the resultant force, $m$ is the mass, and $a$ is the acceleration.

One **newton** (1 N) is the force that will accelerate a mass of one kilogram (1 kg) by one metre per second per second (1 m s⁻²).

The newton has the fundamental units kg m s⁻². As always, it is important to use a consistent set of units in calculations.

### Key term

**Weight** is the gravitational pull that the Earth exerts on an object.

Newton's second law of motion (Newton 2) helps here. Objects accelerate downwards near the surface in the Earth's gravity at 9.81 m s⁻². So Newton 2 indicates that the force $F$ acting on the object must be equal to the mass $m$ multiplied by $a$.

### DP link

You will learn how the gravitational force depends on mass and distance between objects when you study **6.2 Newton's law of gravitation** and also **10.2 Fields at work** (higher level only).

**Inertial or gravitational mass?**

It is possible to think of mass in two ways:

i)  as the response of an object to the application of a force (in other words the smaller the acceleration from a standard force, the larger the mass because $m \propto \dfrac{1}{a}$), or

ii) as the response of an object to the gravity field.

These two descriptions are not the same; they are both physically and philosophically different.

Physicists assume that 1 kg of gravitational mass is equivalent to 1 kg of inertial mass.

## Practical skills: Verifying Newton's second law

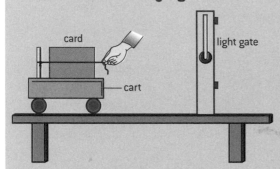

To verify the equation $F = ma$ it is necessary to carry out two experiments: one showing that $F \propto a$ while keeping $m$ constant, and another showing that $a \propto \dfrac{1}{m}$ while keeping $F$ constant.

The same apparatus can be used for both experiments.

**$F \propto a$**

- The force is applied to the trolley using an elastic thread held at a fixed extension. You can do this by keeping your hand in the same position relative to the trolley as the trolley accelerates.

- The measurement of the acceleration of the trolley can be made in a number of ways, including an ultrasound sensor connected to a computer (programmed to provide a direct readout of the acceleration), or a light gate that can time how long it takes the trolley to pass through (in this case you need to use the *suvat* equations to work out the acceleration).

- Measure the acceleration with one, two and three identical threads of the same initial length in parallel (one, two and three equal forces) all extended by the same amount.

- Plot a graph to show the variation of acceleration with force. You should find that the line is (approximately) straight and through the origin.

- Think carefully about errors and how to eliminate them. The friction at the axles of the trolley is a particular problem. What can you do to the table to eliminate this friction? What would you expect a trolley with no resultant friction acting on it to do (think Newton's first law)?

**$a \propto \dfrac{1}{m}$**

- This time you will keep the force the same for each run.

- Add mass to the trolley and measure the acceleration.

- This time plot a graph of $a$ against $\dfrac{1}{m}$. It should be a straight line again.

**Question**

11 The acceleration due to gravity near the surface of Titan, a moon orbiting Saturn, is 1.3 m s⁻². A spacecraft is sent to Titan. It contains a payload with a mass of 250 kg.

   a) Calculate the weight of the payload on Earth.

   b) Calculate the weight of the payload in outer space.

   c) Calculate the mass of the payload on Titan.

   d) Calculate the weight of the payload on Titan.

## Newton's third law of motion

The concept that underpins this law (which we will call Newton 3; the first law will be Newton 1 and so on) is not trivial, no matter how short the statement of the law. You may find it helpful to re-read the first section of this chapter before going on.

When A exerts a force on B, then B must exert an equal and opposite force on A. On the face of it, this is straightforward. The trick comes in correctly identifying the pair of forces, known as an **action–reaction pair.**

Go back to the ball sitting on the table from figure 11. There are two pairs of forces at work here:

- pair 1: the gravitational pull of the Earth on the ball, and the gravitational pull of the ball on the Earth

- pair 2: the force of ball on the table, and the force of the table on the ball—both arise from the deformation of the one by the presence of the other.

Notice that:

- within the pair, the magnitudes are the same, but the directions are opposite

- for this case where there is equilibrium, the magnitudes are the same for all four forces, but this will not be the case when there is acceleration.

Another example is a ball falling under gravity with no air resistance (figure 15).

The Earth exerts a pull on the ball (figure 15). The ball exerts a pull on the Earth of the same magnitude and opposite direction. This is an action–reaction pair (Newton 3).

The Earth's pull on the ball leads to the acceleration (Newton 2) that we identify as *g*. The pull of the ball on the Earth gives rise to an acceleration of the Earth, but this is tiny because $a = \dfrac{F}{m}$ (Newton 2) and *m* is very large. (You may also want to consider what happened to the Earth when the ball was originally moved into the air. Remember: the ball had to be accelerated using a force to move it above the surface.)

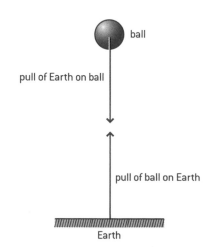

**Figure 15.** An action and reaction pair

**Estimating the acceleration of the Earth**

We can estimate how small such effects on the Earth are. The mass of the Earth is about $6 \times 10^{24}$ kg. What is the effect of a boy jumping down from a wall 1 m high?

As we only need a rough answer, call the mass of the Earth $5 \times 10^{24}$ kg and assume the boy has a mass of 50 kg.

Because the force on the boy and the force on the Earth are the same, $m_{Earth} \times a_{Earth} = m_{boy} \times a_{boy}$ and therefore

$\dfrac{a_{Earth}}{a_{boy}} = \dfrac{m_{boy}}{m_{Earth}} = \dfrac{5 \times 10}{5 \times 10^{24}} = 10^{-23}$. In other words, the acceleration of

the Earth due to the boy is only about $10^{-22}$ m s$^{-2}$.

The skill of making estimates will be an important one for you in both your theory classes and your practical work in DP Physics.

**Are Newton's laws really laws?**

In science, the words "theory", "law" and "hypothesis" have a particular meaning that is more defined than in everyday language.

A theory is a model of some part of the universe. It can use facts, laws and hypotheses. A theory can be used to make a prediction that can be tested by experiment. Theories are often based on earlier theories.

Laws reflect observed patterns of behaviour and often take a mathematical form in physics. They usually do not attempt to explain an effect, but simply state what always happens.

A hypothesis is a possible explanation about the world that may or may not be true. Hypotheses can be tested by experiment and rejected if they prove incorrect.

## Maths skills: Resolving vectors

To **resolve** a vector means to identify two vectors that add to give the original vector. For the IB Physics Diploma Programme this is limited to two vectors that are at 90° to each other. You will need to be able to do this to work out the effect of gravity on objects that are falling while also moving horizontally.

An example is the initial velocity of an object that is thrown into the air at an angle $\theta$ to the ground.

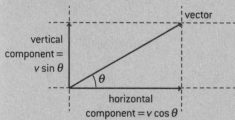

The vector can be resolved either by drawing or algebraically.

For the scale drawing method, draw the original vector and the two directions along which we want to resolve it (usually horizontal and vertical lines) beginning at the start of the vector. Then draw lines from the end of the vector parallel to these directions. These lines intersect with the direction lines at the ends of the two resolved components of the vector.

Algebraically, when the angle between the horizontal and the original vector is $\theta$, then, by trigonometry, one vector is $v \cos\theta$ in the horizontal direction and the other is $v \sin\theta$ vertically.

## Friction effects

In the real world, no surfaces are completely frictionless, the wind blows and air resistance acts. How do we take account of these in calculating motion?

At a simple level friction can be divided into two types: solid friction and fluid friction.

**Solid friction** is the friction observed when one solid surface is dragged over another. Take the case of a book being pulled across a table. Friction arises because the atoms in the book cover interact with the atoms in the table top. Change the materials and the amount of friction will change.

**Fluid friction** is the drag on objects when they move in gases (air resistance) or liquids (viscosity).

Not all friction is wasteful. We make good use of friction in many of the devices we use all the time. Imagine walking to school in a world with no friction!

When we consider friction forces, we encounter examples where there is more than one force acting on a moving object and the forces act in different directions. We need to consider how to deal with the addition of vectors when they are not at right angles to each other. This is best done using an example.

> ### DP link
>
> You will learn about friction and treating it quantitatively when you study **2.2 Forces**. You may also meet fluid friction if you study **Option B.3 Fluids and fluid dynamics**.

> ### Internal link
>
> You will find a description of the phases of matter in **3.1 States of matter**.

### Worked example: Forces acting at angles

6. A girl is dragging a box across rough ground using a rope. The rope is angled upwards at an angle $\theta$ to the horizontal. What force does she need to exert so that the box moves horizontally at a constant velocity?

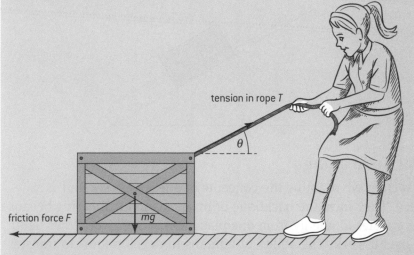

*Solution*

According to Newton 1, because there is no change in velocity there must be no resultant force. Therefore in the horizontal direction the force to the left must be equal to the horizontal component of the tension in the rope, and in the vertical direction the weight downwards must be equal to the vertical component of the tension in the rope. Resolving the forces horizontally shows that $F = T\cos\theta$. Resolving the forces vertically shows that $mg = T\sin\theta$.

We can take the solution in Worked example 6 one step further and divide these equations to show that $\dfrac{T\sin\theta}{T\cos\theta} = \dfrac{mg}{F} = \tan\theta$.

## Maths skills: Trigonometry basics

### Sine, cosine and tangent

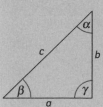

When $\gamma$ is 90° (a right angle), $\sin \beta = \dfrac{b}{c}$, $\cos \beta = \dfrac{a}{c}$, $\tan \beta = \dfrac{b}{a}$.

### Pythagoras's theorem

When $\gamma$ is 90°, $c^2 = a^2 + b^2$

### Sine and cosine rules

For any values of angles $\alpha$, $\beta$, $\gamma$:

$$\frac{a}{\sin \alpha} = \frac{b}{\sin \beta} = \frac{c}{\sin \gamma} \quad \text{(the sine rule)}$$

and $c^2 = a^2 + b^2 - 2ab \cos \gamma$ (the cosine rule).

## Question

12 A block of mass 4.5 kg slides down a ramp at an angle of 30° with a constant acceleration. It travels a distance of 2.5 m from rest in 5.0 s.

    a) Calculate the acceleration of the block.

    b) Calculate the frictional force that opposes the motion of the block.

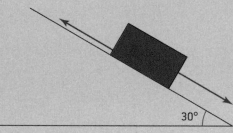

### Internal link

You will gain an insight into how humans use elements of the natural world to generate electrical energy in **6 Generating and using energy**.

### DP link

You will learn about energy and work done when you study **2.3 Work, energy and power**.

## 1.3 Work and energy

In this section we examine the concepts of energy, power and efficiency. They form the backbone of physics and engineering because it is the transfer of energy from one place to another that allows us to extract useful work. The ability of humans to develop tools to transfer energy from one form to another has made us an adaptable and resourceful species.

### Work done

Physicists have a precise, clear meaning for work: work is done when a force leads to the movement of an object. This leads to a definition of *work done* and a unit for energy, the *joule*.

## Worked example: Calculating the work done

7. A force of 15 N acts on a mass and moves it against friction through a distance of 25 m in the direction of the force. Find the work done.

**Solution**

$W = F \times s = 15$ (N) $\times 25$ (m) $= 375$ J.

8. A railway truck moves on rails that are laid west to east. A force of 1.5 kN acts on the truck and it moves 52 m.

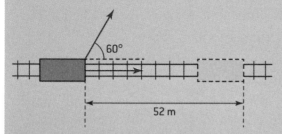

Calculate the work done if the force acts:

**a)** along the direction of travel of the track

**b)** at an angle of 60° to the track.

**Solution**

**a)** Work done $= F \times s = 1500$ (N) $\times 52$ (m) $= 7.8 \times 10^4$ J or 78 kJ

**b)** The component of force in the direction of motion is $F \cos 60$, which is $1500 \times \cos 60 = 750$ N.

So the work done is $750 \times 52 = 39\,000$ J (39 kJ)

 **Key term**

**Work done** = force × distance moved in the direction of the force

$W = F \times s$

The unit of work is the **joule** (abbreviated J). One joule of work (1 J) is done when a force of 1 newton moves an object through 1 metre in the direction of the force.

The joule is not one of the fundamental SI units. From the equation, you can see that it can also be written as newton metres (N m), and this can be further written as kg m s$^{-2}$ × m, in other words kg m$^2$ s$^{-2}$.

Notice that energy is a scalar quantity; it does not have a direction associated with it.

## Maths skills: Using standard form in answers to questions

It is best to write the answer to part (a) of Worked example 8 in the form 75 kJ or $7.5 \times 10^4$ J because using 75 000 J could be taken to mean that you know your answer accurately to 5 significant figures. As the data in the calculation were to 2 sf it is wrong to imply this 5 sf level of accuracy.

## Question

13 Calculate the work done when a force of 12 N moves an object through a distance of

a) 6.0 m

b) 6.0 m at 45° to the direction of the force.

## Energy stores and pathways

Where does the energy to do work come from? Physicists use the concept of the "energy store". Energy is available for use as work when it moves from one store to another. Sometimes the origin of the energy is called the "source" and the store to which the energy goes is called the "sink". Here are some examples.

When the north poles of two magnets face each other, as you try to push the magnets together you store energy in the magnet system. Release the magnets and they fly apart. This repulsion could be harnessed to obtain work (two magnets, one on a railway truck and one on a fixed part of the track, could move the truck).

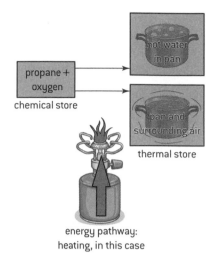

energy pathway:
heating, in this case

**Figure 16.** Energy transfer with a camping stove

### Internal link

Many of these energy stores are important for the generation and storage of energy:

- electrostatic and magnetic – **2 Electric charge at work**
- nuclear – **6 Generating and using energy**
- thermal – **3 Thermal physics**
- global resources – **6 Generating and using energy**.

Numerical ways to represent energy stores and pathways are explored in **6.2 Energy resources**, where there is a discussion of Sankey diagrams.

### Key term

The **principle of conservation of energy** states that energy cannot be created or destroyed.

A spring or a rubber band has no ability to do work when unstretched, but stretch it in your fingers and it can release its store of energy as a catapult.

A camping stove contains propane fuel. When this gas mixes with oxygen from the air, the chemical bonds in the two gases are altered, releasing energy in the stove flame (figure 16). Often people refer to the propane as the store, but it cannot release energy unless oxygen is present, so it may be more correct to regard both gases as the store.

There are many types of energy stores:

- chemical (for example, the propane stove described above)
- elastic (for example, the spring described above)
- electrostatic (energy stored in a system of two electric charges that attract or repel when released)
- gravitational (energy stored in a system of two masses that are attracted by gravity where one can be allowed to move relative to the other)
- kinetic (energy stored in a moving object)
- magnetic (for example, the magnets described before)
- nuclear (energy stored in atomic nuclei, transferred by radioactive decay, nuclear fission or nuclear fusion)
- thermal (energy stored in a hot object).

The principal energy pathways are:

- electrical (a charge moving through a potential difference)
- heating (when there is a difference of temperature)
- mechanical (a force moving an object through a distance)
- radiation (typical wave motion, eg light/radio/sound waves).

### Conservation of energy

When energy transfers from one store to another, observations suggest that none is lost provided we are very careful to include every possible form of energy in our measurements. This is known as the *principle of conservation of energy*.

**DP ready** | **Nature of science**

**Conservation laws**

Some physical quantities are always conserved. These laws are of great importance in the philosophy of the subject as well as in calculations. The laws include:

- conservation of charge
- conservation of linear and angular momentum
- conservation of energy (with the proviso that all energy forms must be considered).

Similarly, there are some fundamental constants that are thought to never change (the charge on the electron is an example).

Physics underwent a **paradigm shift** at the beginning of the twentieth century when Einstein suggested that mass was itself a form of energy, in his famous equation $\Delta E = c^2 \Delta m$. $\Delta$ stands for "change in" and $c$ is the speed of light in a vacuum. This equivalence means that the principles of conservation of mass and conservation of energy can be combined.

**Paradigm shifts**

From time to time in science a discovery or suggestion is made that is so different from the previous view that it causes a major shift in our models of the universe. Examples include:

- Galileo Galilei's observation of Jupiter's moons, which provoked his support for the Copernican view of the solar system
- Isaac Newton's work on the implications for the gravitational force
- Albert Einstein's recognition that time was not an absolute quantity but depends on the conditions of the observer.

**Figure 17.** Galileo Galilei (1564–1642), Isaac Newton (1642–1726) and Albert Einstein (1879–1955)

These are topics you are likely to discuss not only in your Diploma physics lessons but also in theory of knowledge.

**Worked example: Drawing energy stores and energy pathways**

9. An electric cell is connected to a motor that is raising a load. Draw the energy stores and the energy pathways for this process.

*Solution*

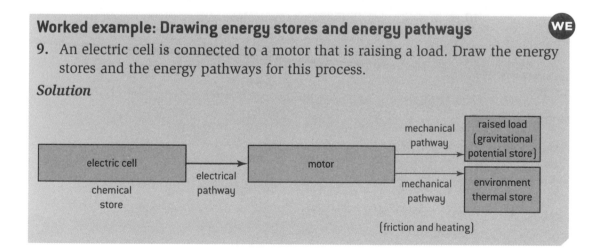

## Kinetic energy and work

In a petrol-driven car, the chemical store of energy consists of the liquid fuel + oxygen in the air. It supplies energy that is eventually transferred into thermal stores, including raised temperature of the friction brakes when the car stops, heated air from air resistance, hot tyres from friction, and so on. Such stores probably cannot be used again, so this energy is "lost" to us. However, while the car is moving the energy that it has is called its *kinetic energy*.

The kinetic energy of an object increases when:

- the speed of the object increases for a given mass
- the mass of the object increases for a given speed.

**Internal link**

Energy pathways in a resource context are discussed in **6.2 Energy resources**.

Suppose that the car and passengers with a total mass $m$ change speed from an initial speed $u$ to a final speed $v$. This acceleration $a$ takes a time $t$ and occurs in a distance $s$.

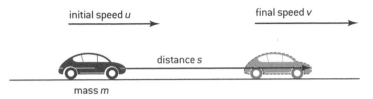

So the work done $W$ is $W = \text{force} \times s = m \times a \times s$ ($F = ma$ has been used here to substitute for $F$).

However, $s$ can also be replaced using the second kinematic equation ($v^2 = u^2 + 2as$) to give

$$W = m \times a \times \left(\frac{v^2 - u^2}{2a}\right)$$

Cancelling the $a$ and rearranging gives

$$W = \frac{1}{2} mv^2 - \frac{1}{2} mu^2$$

The symbol used for kinetic energy in DP Physics is $E_k$.

The kinetic energy of an object of mass $m$ moving at a speed $v$ is $\frac{1}{2} \times \text{mass of the object} \times \text{its speed}^2$.

This confirms the two predictions of how kinetic energy will vary with speed and mass listed on page 23.

listed on page 23.

**Worked example: Calculating change in kinetic energy**

10. A bus of mass 10 000 kg accelerates from a speed of 10 m s$^{-1}$ to a speed of 15 m s$^{-1}$. Calculate the change in the kinetic energy of the bus.

*Solution*

The change in kinetic energy is $\frac{1}{2} \times m \times (v^2 - u^2) =$

$\frac{1}{2} \times 10^4 \times (15^2 - 10^2) = \frac{1}{2} \times 10^4 \times (225 - 100) = 6.3 \times 10^5$ J

Notice that $(15^2 - 10^2)$ is **not** the same as $(15 - 10)^2$. This is a common error: one is 125, the other is 25!

**Question**

14 Calculate the kinetic energy of:

    a)   a tennis ball of mass 58 g moving with a speed of 65 m s$^{-1}$

    b)   a boy on his bicycle of total mass 75 kg moving with a speed of 8.5 m s$^{-1}$.

15 Estimate the kinetic energy of:

    a)   a girl jumping off a wall of waist height when she reaches the ground

    b)   a car being driven at the speed limit

    c)   a bird in normal flight.

**Making estimates**

One of the command terms used in IB Physics Diploma Programme questions is "estimate". When you are asked to estimate, you may be expected to "guess" some, or all, of the quantities involved in the problem, as in the estimate of the acceleration of the Earth in *1.2 Pushes and pulls*. The point of an estimate is to give a power-of-ten value for the answer, not an exact answer.

Outside examinations, use the internet to research reasonable values of quantities you do not know.

It is sensible to use estimates for physical constants. Use $10 \text{ m s}^{-2}$ for $g$ rather than $9.81 \text{ m s}^{-2}$ if you just need a power-of-ten answer.

## Gravitational potential energy and work

The term "potential" in physics requires care. It also occurs in the term "potential difference" in electrical theory (often abbreviated to pd). Some books emphasize that elastic energy (in a compressed spring, for example) is elastic **potential** energy in the sense that the spring has not yet been released to make its energy available.

*Gravitational potential energy* (gpe, with the symbol $E_p$) is energy stored in a system formed by the gravitational interaction of two (or more) objects. All massive objects (meaning objects with mass, not necessarily very big ones) attract each other by gravity and will therefore move towards each other if they can. (Gravitational repulsion is never observed.) When the two objects are held apart, they have the *potential* to move back together, and the energy released from the system can be transferred into useful work. A good example is a hydroelectric power station: water is allowed to flow downwards through a turbine. The turbine gains kinetic energy (it rotates), transferred to it from the kinetic energy of the water that in turn comes from the water's store of gravitational potential energy.

In this and other examples of transfer of gravitational potential energy (gpe), you should carefully analyse the sources and endpoints of energy and the transfer pathways for the energy. As with kinetic energy, your thinking about gpe should be quantitative as well as qualitative. We begin with the definition of work done and extend it to the case of a mass just above the Earth's surface (figure 18).

The mass is a distance $h$ above the Earth's surface. Suppose that the acceleration due to gravity over this distance is effectively constant (it decreases as we move away from the surface but we can ignore this for small height changes). The work done is force × distance, as usual, so in this case work done = weight × height, which is $mg \times h$

so change in gpe = mass × acceleration due to gravity × vertical distance change.

When the object moves away from the Earth's surface the change in gpe is positive; when the movement is towards the surface the change is negative and the object loses gpe.

**DP link**

You will study kinetic energy and gravitational potential energy in **2.3 Work, energy and power**.

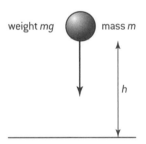

weight $mg$     mass $m$

$h$

**Figure 18.** Gravitational potential energy of a mass above the Earth's surface

**Units of gravitational potential energy**

Gravitational potential energy is weight × height, so this is a force × distance expression and so is equivalent to work done in joules.

## Key term

Changes in **kinetic energy** and **gravitational potential energy**

$$\Delta E_k = \frac{1}{2} m (v^2 - u^2)$$

$$\Delta E_p = mg\Delta h$$

where $m$ is mass, $v$ and $u$ are final and initial speed, $g$ is the acceleration due to gravity and $\Delta h$ is the change in vertical height. Remember that $\Delta$ means "change in".

## DP link

Later in the IB Physics Diploma Programme (**10.1 Describing fields**) you will learn to distinguish between "gravitational potential" and "gravitational potential energy". The former is the energy stored in the two-object system *per unit mass* and is quite different from gpe.

## Worked example: Kinetic and gravitational potential energy calculations

11. A stone of mass 0.50 kg is dropped from rest from a point 2.5 m above the ground. Air resistance is negligible. Calculate the speed of the stone when it hits the ground. Take $g = 9.81$ m s$^{-2}$.

*Solution*

This problem can be solved in two ways: either by using energy ideas or by using the *suvat* equations from *1.1 Faster and faster*. The equations can be used here because the acceleration is uniform. The solution shown uses energy ideas.

The loss of gpe when the stone falls to the ground is
$mg\Delta h = 0.50 \times 9.81 \times 2.5 = 12.26$ J

This *loss* of gpe must equal the *gain* in kinetic energy so

$\frac{1}{2} mv^2 = 12.26$ J. So $v = \sqrt{\dfrac{12.26 \times 2}{0.50}} = 7.0$ m s$^{-1}$.

(Notice that you do not even need the value for mass here. Equating the two energies gives $\frac{1}{2} mv^2 = mgh$ so the mass $m$ cancels to give $\frac{1}{2} v^2 = gh$ and therefore $v = \sqrt{2gh}$.)

## Question

16 A stone has a mass of 3.5 kg. Assume $g = 9.81$ m s$^{-2}$. Calculate for the stone:

a) the gravitational potential energy as it is released from rest from a cliff 7.8 m above the sea surface

b) the kinetic energy when it has fallen vertically through 3.6 m

c) the kinetic energy just before it hits the sea.

## Key term

$$\text{power} = \frac{\text{energy transferred}}{\text{time taken for the transfer}}$$

The unit of power is the **watt** (symbol W). 1 watt $\equiv$ 1 joule per second (1 J s$^{-1}$).

1 W is therefore also 1 N m s$^{-1}$, which in fundamental (SI) units is kg m s$^{-2}$ $\times$ m $\times$ s$^{-1}$ or kg m$^2$ s$^{-3}$.

The watt is a small unit so you should become familiar with expressing power in kW (in domestic situations) and MW or GW (for energy generation on a local or national scale).

## Power

Two cyclists, each with the same total mass of 95 kg, ride up the same hill. The hill is 55 m high. This means that they both transfer 51 kJ ($mgh = 95 \times 9.81 \times 55$) to their store of gravitational potential energy. Cyclist A takes five minutes to climb the hill and cyclist B takes three minutes longer. A is transferring energy into the gravitational form faster than B; we say that A is more powerful than B. *Power* is the rate at which energy is transferred. In this case, cyclist A transfers energy to the gravitational field at a *rate of* $\dfrac{51\,000}{5 \times 60} = 170$ W;

B does so at 106 W $\left(\dfrac{51\,000}{8 \times 60}\right)$.

Power can be expressed in another way.

The expression for work done is force $\times$ distance. So power must be $\dfrac{\text{force} \times \text{distance}}{\text{time}}$. But this is also force $\times \dfrac{\text{distance}}{\text{time}}$, in other words force $\times$ speed. In symbols, power $P$ can be expressed as $P = F \times v$.

## Worked example: Calculating power

12. A lift door of width 1.2 m requires a force of 250 N to open it. An electric motor opens the door in a time of 6.0 s. Calculate the power that the motor must deliver.

*Solution*

The speed at which the door opens is $\frac{1.2}{6.0} = 0.20$ m s$^{-1}$. The power delivered by the motor must be

$250 \times 0.20 = 50$ W.

## Efficiency

The two cyclists described before transferred energy at different rates, but that was not the whole story. We calculated only the rate at which energy transferred into the gravitational potential energy store. There are many other ways in which energy transferred by the rider is lost: to air resistance, to rolling resistance of the tyres, to the chain connecting the pedals to the rear wheel, in the gear train, and in the muscles of the rider. All these factors reduce the *efficiency* of the system. Efficiency is the ratio of the useful work done by a system to the total energy transferred in all forms.

## Worked example: Calculating efficiency

13. An athlete on an exercise bicycle pedals against a resistance force in the bicycle of 25 N. Her speed on the bicycle is equivalent to a ground speed of 12 m s$^{-1}$.

    **a)** Calculate the rate of transfer of energy to the bicycle by the athlete.

    **b)** The athlete's muscles have an efficiency of 20%. Calculate the rate at which energy is supplied to her muscles by her body.

*Solution*

**a)** $P = Fv = 25 \times 12 = 300$ W ($3.0 \times 10^2$ W is better, to get the significant figures to match).

**b)** She outputs 0.3 kW of energy each second. This is 20% $\left(\frac{1}{5}\right)$ of the chemical energy required from her body, so her body supplies $0.3 \times 5 = 1.5$ kW (or 1.5 kJ every second) to her muscles.

## Maths skills: Powers of ten

- Powers of ten are an effective way to avoid using large numbers of zeroes: $3 \times 10^8$ m s$^{-1}$ is a better way to express the speed of light than 300 000 000 m s$^{-1}$.

- Powers of ten allow you to manipulate large and small numbers with ease.

  A thermal power station produces $2 \times 10^8$ W of electrical power. When 1 kg of fuel is burnt it produces $8 \times 10^7$ J of energy. The efficiency of the station is 20%.

  Calculate the mass of fuel burnt in one second.

  The station requires $2 \times 10^8 \times \frac{100}{20} = 10^9$ J of energy to be supplied each second.

### DP link

You will learn about power and efficiency when you study **2.3 Work, energy and power**.

### Key term

You will often see the phrase **rate of** ... used. "Rate of transfer of energy" is an example in Worked example 13. This means the energy supplied per second.

So power is the rate of transfer of energy because it is the energy transferred in one second.

### Key term

**Efficiency** can be defined in terms of either the energy transferred or the power input and output:

$$\text{efficiency} = \frac{\text{useful work done}}{\text{total energy transferred}}$$

$$\text{or} \quad \frac{\text{useful power output}}{\text{total power input}}$$

Efficiency has no units because it is a ratio of two identical quantities. Efficiency can be expressed as a fraction, as a decimal number between 0 and 1, or as a percentage.

## Maths skills: Rules for manipulating powers of ten (indices)

$10^0 = 1$

$10^m \times 10^n = 10^{m+n}$

$10^m \div 10^n = 10^{m-n}$

$(10^m)^n = 10^{m \times n}$

The mass of fuel $= \dfrac{\text{energy required each second}}{\text{energy from one kilogram}} = \dfrac{1 \times 10^9}{8 \times 10^7} = 13$ kg in each second.

- Science has a list of **prefixes** that can be added in front of units to avoid even the powers of ten. The full list is:

| Prefix | Unit | Value | |
|---|---|---|---|
| peta | P | $10^{15}$ | |
| tera | T | $10^{12}$ | |
| giga | G | $10^9$ | billion |
| mega | M | $10^6$ | million |
| kilo | k | $10^3$ | thousand |
| *hecto* | *h* | $10^2$ | *hundred* |
| *deca* | *da* | $10^1$ | *ten* |
| | | | |
| *deci* | *d* | $10^{-1}$ | *tenth* |
| centi | c | $10^{-2}$ | hundredth |
| milli | m | $10^{-3}$ | thousandth |
| micro | μ | $10^{-6}$ | millionth |
| nano | n | $10^{-9}$ | billionth |
| pico | p | $10^{-12}$ | |
| femto | f | $10^{-15}$ | |

*Italics* represent prefixes in the SI system that are seldom used.

### Question

17  A 75 W electric motor raises a mass of 2.5 kg through a height of 1.8 m in 8.0 s.

Calculate:

a)   the electrical energy supplied to the motor

b)   the gravitational potential energy gained by the load

c)   the efficiency of the motor.

## 1.4 Momentum and impulse

### DP link

You will learn about momentum and impulse when you study **2.4 Momentum and impulse**.

Up until now, Newton 2 has been expressed as $F = ma$. This can be rewritten in terms of change in velocity:

$$F = m \times \frac{\text{change in velocity}}{\text{time taken for change}} = m \times \frac{\Delta v}{\Delta t}.$$

$\Delta$ stands for "change in ...". So $\Delta v$ is the change in velocity and $\Delta t$ is the change in time. Another rearrangement of the same equation yields: $F \times \Delta t = m \times \Delta v$ and something interesting appears. Think about the case of an object falling to the Earth. Because there is an action–reaction pair, Newton 3 tells us that the forces acting on the object and the Earth are the same in magnitude ($F$). The two forces must obviously act for the same time ($\Delta t$) and therefore everything on the left-hand side of the equation is the same for both bodies. This means that the quantity (mass × change in velocity) must also be the same for both bodies. This new quantity is of such importance in science that it is given its own name: *impulse*.

It leads us to another new, important quantity: *momentum*. Momentum is the product of (mass × instantaneous velocity) for an object.

## Conservation of momentum

The total momentum of a system remains constant providing no external forces act on it.

This law comes in two parts, and it is important not to overlook the second part, "providing no external forces act", because when a force outside a system acts on it, then the momentum can, and usually does, change.

When two railway trucks collide then at the moment of collision the momentum is unchanged, but as friction in the form of air resistance begins to act on both trucks, then the momentum of the system gradually decreases as energy is transferred into the air.

Momentum is one of the quantities in science that is **always** conserved. The total momentum is never lost, whatever the nature of the interaction. If you think some may have disappeared, you are not looking carefully enough at the problem.

**Key term**

Momentum = mass × velocity

The units of momentum are kg m s⁻¹.

The symbol $p$ is often used for momentum and $\Delta p$ for momentum change.

Momentum is a vector quantity; always specify its magnitude and direction.

Impulse = force × time for which it acts

Impulse = change in momentum

**Internal link**

You meet conservation laws a number of times in this book, including conservation of charge (**2.1 Electric fields and currents**) and conservation of momentum (**3.2 Gas laws**, discussing the motion of gas particles).

---

## Worked example: Conservation of momentum in collisions    **WE**

14. Figure 19 shows two trolleys approaching each other. After colliding, they stick together.
    Calculate the final velocity of the combined trolleys immediately after the collision.

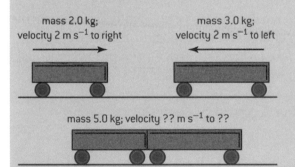

**Figure 19.** Trolley problem

*Solution*

The initial momentum (treating movement to the right as positive) =

$m_1 u_1 = m_2 u_2 = 2 \times (+2) + 3 \times (-2) = 4 - 6 = -2$ kg m s⁻¹

(this means the total momentum is 2 kg m s⁻¹ to the **left**).

After the collision, the momentum must be unchanged, but now the mass is 2 + 3 = 5 kg.

The final velocity $= \dfrac{\text{momentum}}{\text{total mass}} = \dfrac{-2}{5} = -0.4$ m s⁻¹. This is negative so the joined trucks are travelling at a speed of 0.4 m s⁻¹ to the left.

**Question**

**Q**

**18** A 12 kg mass travelling to the left at 7.5 m s$^{-1}$ collides with a 3.0 kg mass travelling to the right at 5.0 m s$^{-1}$. The masses stick together when they collide and continue along the original line.

a) Calculate the initial momentum of the 12 kg mass and the 3 kg mass.

b) Calculate the initial total momentum.

c) Calculate the final velocity of both masses.

**DP link**

You will learn about energy changes in collisions in **2.4 Momentum and impulse**.

**Key term**

In an **elastic collision**: momentum is conserved and energy is conserved.

In an **inelastic collision**: momentum is conserved but energy is transferred out of the system.

In a **superelastic collision**: momentum is conserved but energy is transferred into the system.

## Energy changes in collisions

Look again at the example of the two trolleys colliding in Worked example 14 on page 29. Before the collision the two trolleys had a combined kinetic energy of $\frac{1}{2} \times 2 \times 2^2 + \frac{1}{2} \times 3 \times 2^2 = 4 + 6 = 10$ J. Remember that energy is a scalar and so there is no reason to consider the direction of the trucks.

After the collision the combined mass of 5 kg had a speed of 0.4 m s$^{-1}$. So the final kinetic energy was 0.40 J.

When energy is lost in a collision it is known as an *inelastic* collision. When energy is neither gained nor lost the collision is known as *elastic*. Where energy is input to the system during the collision, it is called a *superelastic* collision or explosion.

For the colliding trolleys, although the momentum of the system was conserved, the kinetic energy was not. Where has it gone? The answer is that the kinetic energy has transferred into several energy sinks. When the trucks collided, energy was used to operate the coupling that joins them together, sound energy was transferred to the air and parts of the truck mechanisms may have deformed, permanently or temporarily. All these changes require an energy transfer. Most or all of this energy will eventually find its way into a thermal (heat) form and become lost to us. For example, the energy transferred into sound waves will be dissipated in the air or nearby objects, leading to very small increases in temperature, which we cannot access to do useful work.

## Worked example: Conservation of kinetic energy in collisions

**WE**

15. A stone of mass 5.0 kg slides across ice on a pond and collides elastically at a speed of 0.20 m s$^{-1}$ with a stone also of mass 5.0 kg. Calculate the final speeds of both stones after the collision.

*Solution*

Initial momentum of the stone is $p = mv = 5.0 \times 0.20 = 1.0$ kg m s$^{-1}$

Initial kinetic energy of system $E_k = \frac{1}{2} mv^2 = \frac{1}{2} \times 5.0 \times 0.20^2 = 0.10$ J

If after the collision the original stone has mass $m_1$ and speed $v_1$, there are two equations, one for final momentum and one for final kinetic energy:

$1.0 = m_1 v_1 + m_2 v_2$

$0.1 = \frac{1}{2} m_1 v_1^2 + \frac{1}{2} m_2 v_2^2$

The solution of these equations is that either $v_1 = 0$ and $v_2 = 0.2$ m s$^{-1}$ or that $v_2 = 0$ and $v_1 = 0.2$ m s$^{-1}$. The second solution can be discounted (it is as though there were no interaction). The first solution says that the final speed of the second stone is the same as the original speed of the first stone. The first stone stops completely, and the second stone moves off in the same direction at the original speed. This is an elastic collision as required.

## Question

Q

**19** Railway truck A has a mass of 800 kg and railway truck B has a mass of 1600 kg. Truck A travels towards truck B at a speed of 8.0 m s⁻¹. Truck B is initially stationary. The trucks join during the collision.

   **a)** Calculate the speed of the trucks immediately after the collision.

   **b)** Calculate the total kinetic energy lost during the collision.

**20** A hammer of mass 5.0 kg hits a thin vertical piece of wood that has a mass of 0.75 kg. The speed of the hammer is 2.5 m s⁻¹ when it hits the wood, and it does not bounce. The wood is driven 7.5 cm into the ground.

   **a)** Calculate the vertical speed of the wood immediately after the hammer hits it.

   **b)** Determine the average frictional force acting on the wood due to the ground.

## Motion in a circle

Many objects move in a circular path. Obvious examples are the path of the Moon or a satellite around the Earth and of the Earth around the Sun. (Although these orbits are not perfectly circular, they are close enough for our purposes.) Likewise, the path of an object being whirled around in a horizontal circle springs to mind.

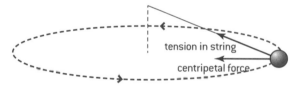

tension in string

centripetal force

**Figure 20.** Centripetal force for an object on a string

Figure 20 shows the situation for an object rotating with a constant speed at the end of a string. What keeps the object rotating? The horizontal component of the tension in the string is providing a force inwards towards the centre of the circle. This force, which is keeping the object in its orbit, is the **centripetal force**.

Looking at all the forces acting on the object leads to figure 21.

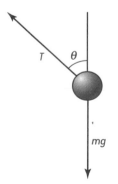

$T$  $\theta$

$mg$

**Figure 21.** The forces acting on the rotating object

The centripetal force is provided by the horizontal component $T \sin \theta$, while the vertical component of the string tension balances the weight force, so $T \cos \theta = mg$. This explains why the string will never be completely

 **DP link**

You will study quantitative details about centripetal motion when you study **6.1 Circular motion**.

horizontal, as some vertical component of *T* always has to compensate for the weight of the rotating object. What determines the size of the component of the force that acts to keep the object rotating in this way?

The centripetal force is increased when:

- the speed of the object or the mass of the object is increased
- the radius of the circle is reduced.

Circular motion is an example of a vector velocity that is **not** constant. The object has a constant speed, but the velocity is always changing (figure 22).

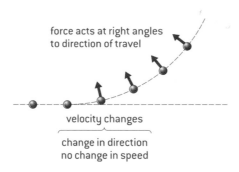

**Figure 22.** The centripetal force acts at right angles to the instantaneous velocity

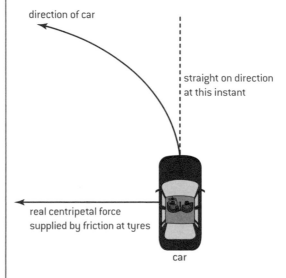

# Chapter summary

Make sure that you have a working knowledge of the following concepts and definitions:

- ☐ The definitions of distance, displacement, speed, velocity and acceleration and the differences between average and instantaneous values of these quantities.
- ☐ Distance–time, speed–time, displacement–time and velocity–time graphs can be interpreted to describe motion.
- ☐ The derivation and use of the *suvat* (kinematic) equations.
- ☐ There are practical methods for making an estimate of the acceleration due to gravity.
- ☐ Balanced forces cancel each other out, leaving a body in equilibrium, and forces add to produce one resultant force.
- ☐ Mass corresponds to the amount of matter in a body, and weight is the gravitational force acting on a body.
- ☐ Newton's three laws of motion are:
  1. every body continues in its state of rest or uniform motion unless external forces act on it
  2. the acceleration $a$ of an object of mass $m$ is related to the force $F$ by $F = ma$ or by $F = m \times \dfrac{\Delta v}{\Delta t}$ where $v$ is the speed of the object and $t$ is the time
  3. every action force has an equal and opposite reaction force.
- ☐ Energy is available for use as work when it transfers between energy stores.
- ☐ Work done = force × distance moved in the direction of the force.
- ☐ Power is the rate of change of energy with time $\left( = \dfrac{\Delta E}{\Delta t} \right)$.
- ☐ Efficiency is $\dfrac{\text{useful work done}}{\text{total energy transferred}}$ or $\dfrac{\text{useful power output}}{\text{total power input}}$.
- ☐ Energy is transferred from an energy store through an energy pathway to an energy sink.
- ☐ Energy is conserved when all final forms, including mass, are taken into account.
- ☐ The kinetic energy $E_k$ of an object of mass $m$ moving at speed $v$ is $\frac{1}{2} mv^2$.
- ☐ The change in gravitational potential energy $\Delta E_p$ of an object of mass $m$ is $+mgh$ when the object is raised through a distance $h$ in a gravitational field of strength (acceleration due to gravity) $g$.
- ☐ Momentum = $mv$ and impulse = $F \times \Delta t$, and momentum is conserved.
- ☐ In inelastic, elastic and superelastic collisions, the system respectively loses, retains and gains energy.
- ☐ For an object to move in a circle, a centripetal force must act on the object towards the centre of the circle, and centripetal force is increased if the speed or mass of the object is increased or if the radius of the circle is decreased.

## Additional questions

1. A motorboat sails with a velocity of 8.0 m s⁻¹ due north. The wind adds a velocity of 6.0 m s⁻¹ due east. Calculate the overall velocity of the boat in the water. Sketch a diagram to show the direction.

2. Two displacements have magnitudes of 18 m and 6 m. Calculate the greatest and least distances of travel that these displacements added together can represent.

3. A force of 9 N and a force of 12 N act on an object. The angle between the forces is 90°.
   a) Determine, using a scale drawing, the resultant of the two forces.
   b) State how it is possible for the two forces to give a resultant of (i) 3 N and (ii) 21 N.

4. A lorry moves from rest with a constant acceleration of 0.35 m s$^{-2}$. The mass of the lorry is 7000 kg.

   a) Calculate the time taken for the lorry to reach a speed of 17 m s$^{-1}$.

   b) Calculate the distance travelled by the lorry in reaching the speed of 17 m s$^{-1}$.

   c) Calculate the initial force required to accelerate the lorry.

   d) In practice, air resistance acts on the lorry, and the magnitude of the resistive force increases with speed. Suggest what this implies for the speed of the lorry.

5. A runner completes a marathon, a distance of 42.2 km, in a time of 3 hours and 30 minutes.

   a) Calculate, in s, the time taken for the runner to complete the marathon.

   b) Determine the average speed of the runner.

6. A car is initially moving at 32 m s$^{-1}$. When the brakes are applied its acceleration is $-4.6$ m s$^{-2}$.

   a) i) Calculate the time taken for the car to stop.

   ii) State the assumption you made in (a)(i).

   b) The mass of the car is 800 kg. Calculate the resultant force acting on the car.

7. Explain the difference between:

   a) energy and power

   b) momentum and impulse

   c) mass and weight.

8.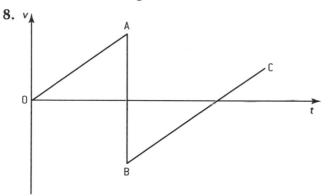

The graph shows the variation of velocity with time for a ball that bounces vertically after release from rest above the ground.

   a) Explain why the gradient of line OA is equal to the gradient of line BC.

   b) Outline why the value of $v$ at B is negative.

   c) Explain why the velocity at B is less than the speed at A.

   d) The ball has a mass of 0.25 kg and is released from 1.5 m above the ground. After the first rebound the ball reaches a height of 1.2 m above the ground.

   Determine:

   i) the speed of the ball immediately before the first impact

   ii) the speed of the ball immediately after the first impact.

9. A motor vehicle of mass 950 kg is claimed to travel a distance of 25 m when it stops on a horizontal road from an initial speed of 18 m s$^{-1}$.

   a) Determine the average deceleration of the vehicle.

   b) Calculate the average frictional force that acts on the vehicle.

10. In a test of a motor vehicle's safety, the vehicle is stopped in a distance of 5.5 m from a speed of 28 m s$^{-1}$. A test dummy of mass 65 kg is wearing a seat belt that allows the dummy to move 0.50 m relative to the vehicle.

   a) Determine the deceleration of the dummy.

   b) Calculate the resultant force that acts on the dummy.

11. An aircraft has a total mass of $3.2 \times 10^5$ kg. It is powered by engines that have a total maximum thrust of 1.1 MN.

    a) Calculate the initial maximum acceleration of the aircraft, ignoring frictional forces.

    The aircraft starts its take-off from rest and has a take-off speed of 95 m s$^{-1}$.

    b) Calculate the time to reach take-off speed, ignoring frictional forces.

    c) In practice, the frictional forces reduce the acceleration of the aircraft to 2.5 m s$^{-2}$. Calculate the mean total frictional force that acts on the aircraft during take-off.

    d) Calculate the minimum length of runway required.

    e) The aircraft travels at a constant velocity and at a constant height after take-off. Explain, with reference to horizontal and vertical forces, how this is achieved.

12. A trolley of mass 60 kg is pushed 25 m at a constant speed up a ramp by a force of 55 N acting in the same direction as the direction of motion. The ramp is 2.0 m high.

    Calculate:

    a) the work done pushing the trolley up the slope

    b) the change in gravitational potential energy of the trolley

    c) the energy wasted in friction.

13. A truck in a fairground ride of total mass 1400 kg moves at an initial speed of 2.0 m s$^{-1}$ before descending through a vertical distance of 50 m to reach a final speed of 28 m s$^{-1}$. The truck travels on a track of length 70 m in this motion.

    a) Calculate the loss of gravitational potential energy of the truck.

    b) Determine the change in kinetic energy during the motion.

    c) Determine the average frictional force on the truck during its descent.

14. A motor vehicle travels along a horizontal road. When the car travels at a constant velocity of 10 m s$^{-1}$, its effective power output is $1.8 \times 10^4$ W. A resistive force acts on the vehicle. This resistive force consists of two components. One is a constant frictional force and is of magnitude 250 N. The other is the air resistance force and is proportional to the car's speed.

    a) Determine the total resistive force acting on the vehicle when travelling at a speed of 10 m s$^{-1}$.

    b) i) Calculate the force of air resistance when the vehicle is travelling at 10 m s$^{-1}$.

       ii) Calculate the force of air resistance when the vehicle is travelling at 5.0 m s$^{-1}$.

    c) Calculate the effective output power of the vehicle when it is moving at a constant speed of 5.0 m s$^{-1}$.

15. Estimate the vertical distance through which a 2.0 kg mass would need to fall to lose the same energy as a 35 W lamp will radiate in 90 s.

# Electric charge at work

> ❝ Magnetic *lines of force* convey a far better and purer idea than the phrase magnetic current or magnetic flood: it avoids the assumption of a current or of two currents and also of fluids or a fluid, yet conveys a full and useful pictorial idea to the mind. ❞
>
> **Michael Faraday, 1854. In Thomas Martin (ed.), *Faraday's Diary: Being the Various Philosophical Notes of Experimental Investigation* (1935), Vol. 6, 315.**

## Chapter context

The **electric currents** created by **moving electrons** are vital to us, from the microscopic currents in our brains to the immense power outputs of modern generating stations. This chapter describes electrical current and explains some of its effects.

## Learning objectives

In this chapter you will learn about:

→ **electric charge** and **electric field**

→ **electric current, potential difference**, energy transfer and **resistance**

→ **Ohm's law** and *I–V* characteristics

→ **series** and **parallel circuits**

→ **permanent magnets** and **magnetic field**

→ **electromagnetic induction**.

## 🔑 Key terms introduced

→ charge carriers
→ electric current
→ coulomb
→ capacity
→ potential difference
→ electromotive force
→ resistance
→ Lenz's law

## ⊗ DP link

You will learn about charge and electric field when you study **5.1 Electric fields** in the IB Physics Diploma Programme

## 2.1 Electric fields and currents

In this section we examine the underlying concepts of electric current: what is meant by charge, and what causes it to flow.

### Charge and electric field

Run a plastic comb through your hair. The comb can then pick up small pieces of paper (figure 1). Rub a balloon on a woollen sweater and the balloon can stick to a vertical wall. We say that the comb and the balloon have become charged—meaning that electrons have transferred between your hair and the comb, or between the sweater fibres and the balloon.

**a)**   **b)**

**Figure 1.** Running a comb through your hair will cause the comb to become charged

These effects rely on electron movement and the fact that charged **objects of the same charge repel** and charged **objects with opposite charge attract** (figure 2).

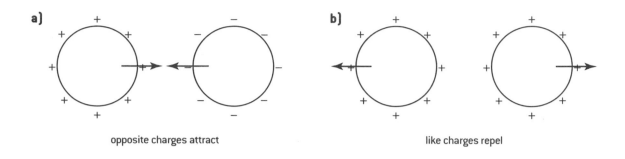

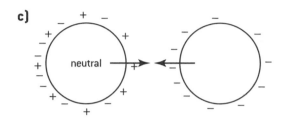

Figure 2. Charge interacting

The comb begins with equal numbers of positive and negative charges. After combing, this is no longer true because electrons have transferred from comb to hair or vice versa, leaving the fixed protons behind. The transfer direction depends on the material from which the comb is made. If the comb has donated electrons to the hair, then the comb will be left with an overall positive charge. This means it will repel other positive charges and can attract either negative charges or neutrally charged objects.

Figure 2 shows the three possibilities that can occur. There is an important **conservation rule**: when object A loses 500 electrons to object B, then object B gains 500 extra electrons.

You may be initially puzzled by the comb picking up pieces of paper that have no overall charge. The positively charged comb causes some free negative charges on the paper to be attracted to the side closer to the comb—the charges do not leave the paper, though. This leaves some atoms on the far side of the paper with a deficit of electrons, so a positive charge. The negative charges in the paper are now closer to the comb than these positive charges and the force on them due to the comb is stronger. The paper is attracted to the comb.

You may also notice that, after it touches the comb, the paper suddenly flies away. Again, this is explained in terms of charge movement. Some negative charges are now transferred to the positive comb, the paper becomes positively charged overall and so is now repelled.

 **Internal link**

You can read more about conservation rules in **1.3 Work and energy**.

**DP ready    Nature of science**

**Positive or negative?**

The story of the development of electrical theory is a fascinating and lengthy one. In short, scientists realized that there were only two types of charge, which they named "positive" and "negative". Assigning the two types meant that the charge on an electron was labelled as negative. There is no particular reason for this; it was just chance. All our electrical rules are based on this assignment and physicists probably will not change it now.

It was the work of JJ Thomson during the late 19th century that led to (more-or-less) direct observation of the electron. All electrons are observed to have the same amount of charge (in other words, are acted on to the same extent by a standard electric field), and we believe that the electronic charge is constant throughout the universe.

**Internal link**

There is more information about the electrons and protons in materials in **5.1 Inside the atom.**

## Electric field and field lines

When an object is charged and has an excess or a deficit of electrons, then it will experience a force when it is in the **electric field** produced by another charge. We explain charge not by what it is, but by its observed effect. In the case of the comb picking up the uncharged paper, we say that the paper is within the electric field of the comb.

> **DP ready**  **Nature of science**
>
> ### Field – a unifying principle
>
> It seems like magic: an invisible force is exerted on a charged object due to something called a **field**. The concept of a field is an important one in science; it helps physicists to draw parallels between different areas of the subject. In the IB Diploma Physics Programme, you will meet electric fields, gravitational fields and magnetic fields. The first two share a mathematical relationship: the strength of the field is proportional to $\dfrac{1}{\text{distance}^2}$. The relationship demonstrates a fundamental and underlying principle about the nature of space itself. For a student, the similarity also means that the two are easier to learn.

Electric fields are visualized using field lines (figure 3). These help us to imagine the field, in a similar way to the magnetic field lines used later in this chapter.

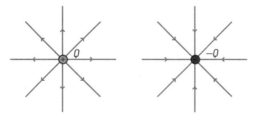

**Figure 3.** Radial field of a single point charge

Figure 3 is a two-dimensional representation of the electric field lines associated with single point charges; imagine lines radiating out (or going in) in all directions with some lines straight out of the page and some into the page. Notice that the field lines are arranged so that they all point to (or away from) the point charge; this is called a *radial field*.

The idea of a field line is that, when a small positive "test" charge is placed on the field line and released, then the test charge will move along the line in the direction of the arrow. A negative charge will

**Internal link**

The concept of electric field lines is very similar to that of magnetic field lines. You will find out more about the properties of field lines in **2.3 Magnetism at work.**

move in the opposite direction. Imagine a positive charge on a line in figure 3 near the –*Q* charge. It ought to be attracted and so the line direction is correct. The arrows on the field lines always show the direction in which a positive charge will move.

To sum up, when a charge is placed in an electric field, a force will be exerted on it because of attraction or repulsion. When the charge is free to move, the force will lead to an acceleration of the charged particle. **An electric force is exerted on any charge placed in an electric field**.

## Charge and electric current

Metals contain free electrons. When the atoms bond together to form a metallic solid these electrons are released from the atoms to which they were originally attached. In the absence of an electric field, these free electrons move around within the metal, colliding with the atoms at high speed but, on average, staying where they are. However, when an electric field acts within the metal, then the electrons experience an electric force and are accelerated. They will begin to move within the metal in the direction dictated by the electric field.

**DP ready    Nature of science**

### Conductors and insulators

Much of the discussion here is concerned with the electrical properties of *conductors*, usually metallic conductors. However, there are other materials with different electrical properties.

Insulators can be imagined as substances that have very few free electrons, so that conduction within the insulator is very difficult. Nevertheless, these properties can be changed: heat glass to a high temperature close to its melting point and it will conduct easily.

Further materials include semiconductors, which have a resistance midway between that of a conductor and an insulator. Semiconductors have unusual properties, some of which are of immense importance in the design of the integrated circuits that make up modern electronic equipment.

The movement of electrons is complicated, but an electron in an electric field can be thought to accelerate, bump into a nearby atom, thus losing some energy, then accelerate again. More precisely, electrons gain kinetic energy through the field and then lose it to an atom through interacting with it. The electron then repeats the process. The net effect is to transfer energy, first from the field to the electron and then from the electron to the atom and, finally, to the metal that is made from the atoms.

This start–stop electron motion evens out and, when an electric field acts on the metal, the electrons move along the metal at a constant average speed. They carry charge, so the charge is also moving at a constant average speed. Such speeds turn out to be quite small in metals, of the order of millimetres per second or even smaller. There are very many electrons per cubic millimetre in a metal, so individual electrons do not have to travel quickly for a large total charge to move through the material.

**DP link**

In the IB Physics Diploma Programme you will study:

- the basic properties of electric fields and how they lead to charge flow in **5.1 Electric fields**

- the theory of both electric and gravitational fields in **10.1 Describing fields** and **10.2 Fields at work**

- some aspects of gravitational fields and their applications to planetary and satellite motion in **6.2 Newton's law of gravitation**.

**Internal link**

We look at the energy that is transferred from the field to the electrons and eventually to the conductor in more detail in **2.2 Electrical resistance**.

**Key term**

A **conductor** is a material containing free electrons or other **charge carriers** so that charge can flow through it when an electric field is applied.

An **electric current** is a flow of electrical charge.

The unit of charge is the **coulomb** (C). This is named for Charles-Augustin de Coulomb, who carried out important experimental work in France during the 18th century.

The coulomb is a large unit in charge terms. The charge of one electron is $-1.6 \times 10^{-19}$ C, meaning that roughly $6 \times 10^{18}$ electrons are required for each coulomb.

 **Key term**

Despite the nature of charge and its key role in the production of an electric current, charge is not a fundamental (base) quantity in SI. Instead, electric current is used as the base unit. Current and its unit, the ampère, are defined later in **2.3 Magnetism at work**.

Charge is defined in terms of **ampères** (amp, A).

**1 C is the charge that flows past a point when a current of 1 A exists for 1 s.**

This flow of charge is called an *electric current*. The electrons in the metal are known as the *charge carriers*. Electrons are not the only charge carriers, and materials other than metals can also carry an electric current.

It is important to understand the conduction model clearly. The chain of argument again:

- An electric field acts on a conductor that contains free electrons.
- The electrons experience a force because of the electric field.
- Through repeated collisions with atoms, the electrons move at a constant average speed in the conductor.
- This movement of charge carriers is what we call an electric current in the conductor.

**DP ready** | **Nature of science**

### Current or charge

People often talk about electric current flowing in a wire. But this is not correct. The electrons are moving or "flowing". They possess the charge and so the charge is "flowing" too.

The wire contains an electric current, or there is a current in the wire – but it is not correct to say that current flows: the current is the flow.

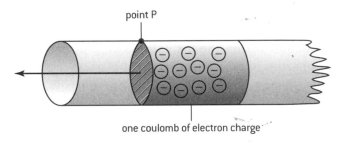

**Figure 4.** Free electrons in a conductor

We can now link charge flow and current in a wire. Figure 4 shows a cylindrical metal conductor containing free electrons (indicated by $\ominus$). These electrons move to the left through the conductor. The relationship between charge $Q$, current $I$ and time $t$ is

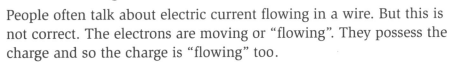

$$\text{electric current} = \frac{\text{total charge that passes a point}}{\text{time taken for the charge to flow past the point}}$$

or $I = \dfrac{Q}{t}$.

Imagine that the shaded volume of the wire contains exactly 1 C of electric charge ($6 \times 10^{18}$ free electrons in this volume will make one coulomb). For the electric current to be one *ampère*, all these electrons must go past point P in one second. At the end of this second, the shaded volume will have moved to the left of P and another 1 C of charge will be about to go past P. Try to keep this picture of the flowing electrons (and charge) in your mind when you study electric current.

### Electronic or conventional current

So far, we have talked about electric current as being due to the flow of free negatively charged electrons in a metal conductor. But as you probably know, current direction is indicated on a circuit diagram in terms of a flow of positive charge. Why is this different from the actual situation?

The answer is largely historical. Although the idea of very small charges is an old one, it was only at the end of the 19th century that scientific observations began to suggest the existence of electrons.

Before that, the convention had been the assumption that positive charge flows in a conductor. To change our notation now would make the rules and descriptions already developed invalid. It is simpler to continue to imagine positive charge flowing, even though everyone knows that it is the opposite in metals.

A current direction that assumes a flow of positive charge is called a *conventional current*; a current direction due to movement of negative charge is an *electron or electronic current*.

Research in the 20th century showed that in fact some semiconducting materials have positive charge carriers.

### Worked example: Calculating charge

1. Calculate the charge that flows when:

    **a)** a current of 10 A flows for 25 s

    **b)** a current of 0.75 A flows for 50 s.

*Solution*

**a)** $Q = It$ so charge $= 10 \times 25 = 250$ C

**b)** charge $Q = 0.75 \times 50 = 38$ C.

### Question

1　A charge of 3.0 C flows past a point in a wire in 500 s. Calculate the current in the wire.

2　Calculate the number of electrons that pass a point in a wire in 10 minutes when the current is:

    **a)** $1.5 \times 10^{-6}$ A

    **b)** 12.5 A.

 **Internal link**

You can find out more about the thermal energy transferred by an electrical conductor in **2.2 Electrical resistance**.

### Energy transfers – defining the volt

As electrons move through a conductor they transfer energy. This energy originates in the electric cell or other device that is connected to the conductor, and at least some, if not all, of it ends up as thermal energy in the conductor. The cell sets up an electric field in the conductor and, as we saw in the last section, this leads to the motion of free electrons.

The obvious way to measure the energy and power being transferred would be to use the joule (see chapter 1). However, this unit is not appropriate in electricity for a very good reason: when an electric cell is connected to a conductor, charge flows but the length of time for which it flows depends on the *capacity* of the cell (how long it takes for the cell to run down). Two cells, one large, one small, can both deliver the same amount of energy per second, but the large cell will do this for longer.

 **Key term**

The **capacity** of a cell is measured in ampère–hour (A h), so a cell with a capacity of 10 A h can in theory deliver 10 A for 1 hour or 100 A for 0.1 h or 10 mA for 1000 h.

The typical capacity of an AAA cell (using the manufacturer's notation for cell size) is 1 A h; for a typical D cell it is 6 A h. Both cells transfer the same energy per second to a conductor connected across them, but the D cell will supply this energy per second for six times longer than the AAA cell.

This total energy transfer by a cell during its lifetime does not help a physicist trying to make a measurement of the energy transfer at one instant. The quantity used for a device in an electrical circuit is the **energy transfer that occurs when one coulomb of charge flows through the device**.

The reason for using this measurement becomes clear when you remember that a coulomb of charge corresponds to a fixed number of electrons flowing through the device. You can imagine a meter that is measuring the energy transfer in the device. As the first electron enters, think of the meter switching on and recording the energy transfer. As the last electron of the coulomb moves out of the device, energy transfer recording is complete.

So, energy transfer in electricity is measured as

$$\frac{\text{energy transferred}}{\text{charge that flows during the transfer}}.$$

The unit of this quantity is J C$^{-1}$, otherwise known as the *volt* (symbol V). When you hear the word "volt" remember that it is a "joule per coulomb". For example, a 1.5 V AA cell transfers 1.5 J for every coulomb that goes through it. This is an effective way to remember that the volt is simply the special unit of electrical energy.

### Pd or emf?

The volt is used not only for cells and power supplies that supply energy **to** electrical circuits, but also for devices like resistors that absorb the transferred energy **from** the circuit. However, two different terms are used to distinguish between these cases.

- When energy is transferred **from** an energy source (such as a cell) to an electrical circuit, the term *electromotive force* (emf) is used.
- When energy is transferred **to** an energy sink (such as a resistor or filament lamp), the term *potential difference* (pd) is used.

Table 1 shows some of the devices used in electrical circuits and indicates the nature of the energy transfer and the appropriate term (emf/pd) in each case.

## Key term

The unit of energy transfer (**potential difference, pd**, or **electromotive force, emf**), $\frac{\text{energy}}{\text{charge}}$, is $1\,\text{V} \equiv 1\,\text{J C}^{-1}$. The **volt** is not a fundamental unit. In base SI units it is written as $1\,\text{kg m}^2\,\text{s}^{-3}\,\text{A}^{-1}$.

## Internal link

**2.5 Practical aspects of electrical physics** is a section of information for you to refer to while working through **2.1 Electric fields and currents** and **2.2 Electrical resistance**.

**Table 1.** Energy sources and sinks in electrical circuits

| Device | | | | | emf or pd? |
|---|---|---|---|---|---|
| cell | | chemical | | electrical | emf |
| resistor | | electrical | | internal (thermal) | pd |
| lamp | transfers energy from | electrical | into | internal (thermal) | pd |
| photovoltaic cell | | Sun (thermal) | | electrical | emf |
| dynamo | | kinetic | | electrical | emf |
| electric motor | | electrical | | kinetic | pd |

### Worked example: Calculating energy transfer

2. Calculate the energy transfer during 1500 s in a component when the pd across it is 12 V and the current is 2.0 A.

*Solution*

The charge that flows in 1500 s is 2.0 × 1500 = 3000 C.

12 J of energy are transferred for every coulomb, so the total energy transfer is 36 kJ.

### Question

3 A filament lamp marked "3 V, 0.6 W" is connected to a 3.0 V battery. Calculate:

  a) the current in the lamp

  b) the energy transfer in the lamp over a time of 1800 s.

4 An electrical appliance works on a mains voltage of 120 V and is rated at a power of 600 W. Calculate:

  a) the energy transferred in 1.0 minute

  b) the current in the device.

## 2.2 Electrical resistance

### Key term

**Resistance**, given the symbol $R$, is defined as

$$resistance = \frac{potential\ difference\ across\ a\ component}{current\ in\ the\ component}$$

Or, in symbols, $R = \dfrac{V}{I}$. This can also be written as $V = IR$ and $I = \dfrac{V}{R}$.

The unit of resistance is the **ohm** ($\Omega$; a Greek capital omega), named for Georg Simon Ohm, a German physicist who lived around the turn of the 19th century. In fundamental units, the ohm is

$$\frac{V}{A} = \frac{(J/C)}{A} = \frac{J}{A\ C} = \frac{kg\ m^2\ s^{-2}}{A \times A\ s} = kg\ m^2\ s^{-3}\ A^{-2}.$$

Clearly, writing $\Omega$ is much easier!

$1\ \Omega$ is the resistance of a component when there is a current of 1 A in it and a potential difference of 1 V across it.

When electrons move in an electrical circuit they act as agents transferring energy from the power supply to the internal energy (thermal energy) of the conductor. Some conductors can transfer energy easily, others poorly so that little energy is transferred as the electrons go through. Metals such as copper, silver and gold are good conductors, so they are poor at transferring energy from electrons. Tungsten is an example of a metal in which the energy transfer is high. Tungsten was used for lamp filaments in incandescent light bulbs.

 **Internal link**

**2.5 Practical aspects of electrical physics** is a section of information for you to refer to while working through **2.1 Electric fields and currents** and **2.2 Electrical resistance**.

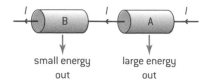

Figure 5. Energy transfer in resistors

Physicists represent these differences using *resistance*. To understand resistance, imagine that a current goes through two conductors (figure 5). Because the conductors are in series, the current in each conductor must be the same. Now imagine that resistor A transfers a great deal of energy in one second but that B only transfers a small amount. During the time interval of 1 s, the same charge flows through each resistor, and therefore the energy transferred per coulomb for A is greater than for B. So the pd across A is larger than for B. From the definition of $R = \dfrac{V}{I}$, the resistance of A is also greater than the resistance of B.

## Practical skills: Determining resistance

Determining the resistance of a metal wire at constant temperature

Table 2. Current, pd and resistance measurements

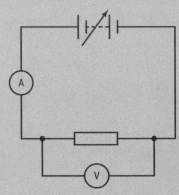

| Current (A) | Pd (V) | Resistance (Ω) |
|-------------|--------|----------------|
| 0.10 | 0.13 | 2.55 |
| 0.20 | 0.26 | |
| 0.30 | 0.76 | |
| 0.40 | 1.01 | |
| 0.50 | 1.27 | |
| 0.60 | 1.53 | |

A suitable circuit for this experiment, using a variable power supply, is shown. Measure the pd across the resistor using a voltmeter and repeat for different currents. Specimen results are shown in table 2. For each pair of values, calculate the resistance; the first has been completed for you. These are real results, so you should expect some variation. Then calculate the average of the resistances.

### Determining the resistance of a lamp filament

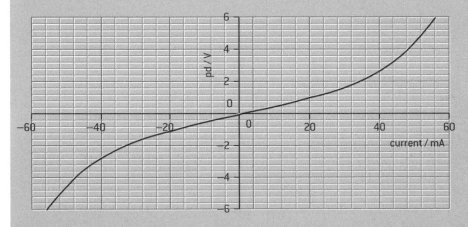

Figure 6. V–I curve for a lamp filament

When the temperature of a conductor varies, so does its resistance (unlike the metal wire in the first experiment). This can be seen from the graph in figure 6, obtained using the same apparatus as above. At a higher pd, more energy is transferred into the filament, and its temperature rises. However, the process of working out the resistance is the same: from the graph read off the current and the pd at a particular current value, then calculate $R = \dfrac{V}{I}$.

Here are the results for two values of current:

At 20 mA, the pd is 1.0 V so $R = \dfrac{1.0}{20 \times 10^{-3}} = 50\ \Omega$;

At 40 mA the pd is 2.8 V so $R = \dfrac{2.8}{40 \times 10^{-3}} = 70\ \Omega$.

The resistance increases as the temperature of the filament increases.

Calculate the resistance of the filament when the current is 30 mA and 56 mA.

Finally, notice that it is **not** correct to calculate the resistance from a *V–I* graph using a tangent at the current position. Resistance is defined as the $\dfrac{V}{I}$ value for a specified current, not an average, which is what working out a tangent or joining the data point to the origin would give.

## Worked example: Pd and resistance calculations

3.  The current in a conductor is $3.0 \times 10^{-3}$ A when the pd across it is 8.0 V. The resistance of the conductor does not change with current.

Calculate:

a)  the resistance of the conductor at $3.0 \times 10^{-3}$ A

b)  the pd across the conductor when the current is $5.0 \times 10^{-6}$ A.

*Solution*

a)  $V = IR$, so $8.0 = 3.0 \times 10^{-3}\,R$; $R = \dfrac{8.0}{3.0 \times 10^{-3}} = 2.7 \times 10^3\ \Omega$

b)  As the resistance remains $2.7 \times 10^3\ \Omega$, the pd is now $5.0 \times 10^{-6} \times 2.7 \times 10^3 = 0.013$ V

## Question

5   Calculate each missing value to complete the table.

| Current (A) | 0.45 | | 1.5 | $6.5 \times 10^{-3}$ | |
|---|---|---|---|---|---|
| pd (V) | 12 | 15 | | 2.5 | 2.5 |
| Resistance (Ω) | | 350 | 18 | | 3700 |

## Ohm's law

Georg Simon Ohm was the first person to recognize that the potential difference in a component is directly proportional to the current in the component providing that the physical conditions of the component (e.g. the temperature) are constant. This is now known as Ohm's law.

Components that obey Ohm's law are said to be **ohmic conductors**. When a component does not obey the law—because the *V–I* graph is curved or not symmetrical—then the component is a **non-ohmic conductor**.

## *I–V* characteristics

Plotting a characteristic curve for any component tells you a great deal about the properties of the component as the current in it changes. Figure 7 on the following page shows the *I–V* **characteristics** for a number of components.

**Internal link**

You can find out more about direct proportion and inverse proportion in **3.2 Gas laws** in the section on gas pressure.

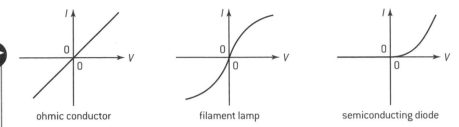

ohmic conductor  filament lamp  semiconducting diode

**Figure 7.** *I–V* characteristics for three components

**What changes resistance?** We have seen that changing the temperature of a resistor can change its resistance. It is easy to see the reason. The atoms in a solid are fixed in position but oscillate (vibrate) around this position. Their degree of movement depends on the temperature of the solid. As the temperature of a conductor rises, the atoms vibrate more and with a greater amplitude. As a result the free electrons moving through the conductor interact more strongly with the atoms and transfer more energy to them. This is equivalent to the resistance increasing (pd increases, current stays the same). This can create a runaway effect so that the increasing energy transfer eventually melts the conductor. This is the effect on which the fuse is based, as a simple way to protect domestic electrical supplies from a current that is too great for the circuit.

It helps to keep the picture of the electron–atom interaction in mind when thinking about other factors that can change the resistance of a conductor.

- **Length:** The longer the conductor, the more atoms any electron is likely to interact with on its way through the conductor, so the greater the resistance.

- **Area:** For a given current, free electrons in a thinner cross-sectional area have a narrower channel in which to move and so they will interact more often on their way through the conductor. So the narrower the channel, the higher the resistance.

- **Material:** Some materials are good conductors, others much worse. Metals such as copper, gold and silver have much lower resistances than metals such as iron and tungsten conductors of the same size.

We will revisit these ideas of change of sample size and shape when we have looked at the combination of resistors in series and parallel.

### Resistances in combination

In a physics or electronics lab, resistors often need to be combined to obtain the required value of resistance.

**Combining resistors in series:** Suppose that three resistors are combined, with values $R_1$, $R_2$ and $R_3$, in series (figure 8) to make one equivalent single resistor with the value $R$.

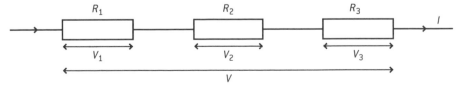

**Figure 8.** Resistors in series

The separate pds across the three resistors are $V_1$, $V_2$ and $V_3$ when the current through each of them is $I$. For a single resistor to be exactly equivalent to the three in series, the current through it must also be $I$. The pd $V$ across the single resistor must lead to the same total energy transfer per coulomb as the three resistors together: so, $V = V_1 + V_2 + V_3$. However, we can now use the definition of resistance to substitute for the pd and therefore $IR = IR_1 + IR_2 + IR_3$. This leads to the rule for combining resistors in series: $R = R_1 + R_2 + R_3$.

**Combining resistors in parallel:** This time imagine the three resistors are in parallel (figure 9).

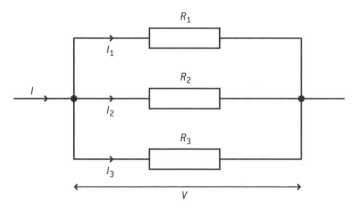

**Figure 9.** Resistors in parallel

When the current in the single resistor (value $R$) is $I$, the sum of the currents in the three resistors must be equal to this, so $I = I_1 + I_2 + I_3$. This time the pd across R and each of the other three resistors is equal to the single value $V$. Using the definition of resistance (this time as $I = \dfrac{V}{R}$), $\dfrac{V}{R} = \dfrac{V}{R_1} + \dfrac{V}{R_2} + \dfrac{V}{R_3}$ and (cancelling $V$) $\dfrac{1}{R} = \dfrac{1}{R_1} + \dfrac{1}{R_2} + \dfrac{1}{R_3}$.

**Internal link**

The theory that the three currents $I_1$, $I_2$ and $I_3$ add to give the current $I$ is covered in **2.5 Practical aspects of electrical physics**.

---

## Practical skills: Testing the resistor combination rules

You need a handful of carbon resistors with different resistance values and a digital resistance meter (or simply a digital multimeter set to measure resistance).

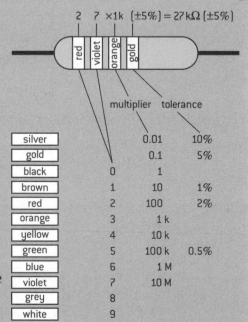

$2 \quad 7 \quad \times 1k \quad (\pm 5\%) = 27\,k\Omega \;(\pm 5\%)$

red | violet | orange | gold

multiplier   tolerance

First, measure the resistance of each resistor. Then choose some series, parallel and series-with-parallel combinations and predict the resistance of each combination. Connect the resistors together electrically and test the combination to see whether your prediction was correct.

| | | multiplier | tolerance |
|---|---|---|---|
| silver | | 0.01 | 10% |
| gold | | 0.1 | 5% |
| black | 0 | 1 | |
| brown | 1 | 10 | 1% |
| red | 2 | 100 | 2% |
| orange | 3 | 1 k | |
| yellow | 4 | 10 k | |
| green | 5 | 100 k | 0.5% |
| blue | 6 | 1 M | |
| violet | 7 | 10 M | |
| grey | 8 | | |
| white | 9 | | |

Resistors are colour coded using a series of rings (figure 10). These bands give you the both the resistance of the resistor and its *tolerance*.

**Figure 10.** Coloured rings signal resistance and tolerance values

**Key term**

**Tolerance** is the percentage uncertainty associated with a value, here the nominal value of the resistance. For example, a gold band at the right-hand end of a resistor means that the value is $\pm 5\%$. So a 1000 $\Omega$ resistor with a gold band can have a resistance somewhere between 950 $\Omega$ and 1050 $\Omega$.

## Maths skills: Combining uncertainties

You will need to take percentage uncertainties into account in combining resistance values in series or parallel.

The uncertainty in a measurement can be expressed in three ways:

- **Absolute uncertainty** – the numerical uncertainty associated with a quantity. For example, when a resistor of quoted value 100 $\Omega$ has an actual value somewhere between 95 and 105 $\Omega$, the absolute uncertainty is $\pm 5$ $\Omega$.

  The resistance is usually expressed as $(100 \pm 5)$ $\Omega$.

- **Fractional uncertainty** – the value of $\dfrac{\text{absolute uncertainty in quantity}}{\text{numerical value of quantity}}$.

  A fractional uncertainty has no unit. For the resistor above, the fractional uncertainty is $\dfrac{5}{100}$ or 0.05.

- **Percentage uncertainty** – the fractional uncertainty $\times$ 100 and therefore expressed as a percentage. Again, there is no unit. For the resistor above the percentage uncertainty is 5%.

When quantities with an uncertainty must be combined mathematically, the result—the **derived quantity**—will also have an uncertainty associated with it. Here are the rules that allow you to calculate the uncertainty in derived quantities:

**When quantities are added or subtracted then the absolute uncertainties are added.**

Two resistors of value $100 \pm 5$ $\Omega$ and $60 \pm 3$ $\Omega$ are connected in series. What is the total resistance of the two in series?

$R = R_1 + R_2$ so the total resistance is 160 $\Omega$.

So in this case the total resistance is $(160 \pm 8)$ $\Omega$.

Expressed algebraically, if $y = a \pm b$, then $\Delta y = \Delta a + \Delta b$; note that when the quantities themselves are subtracted then the uncertainties are still added.

**When quantities are multiplied or divided the fractional uncertainties are added.**

The two sides of a rectangle are $2.0 \pm 0.1$ m and $3.0 \pm 0.3$ m. What is the area of the rectangle?

The area of the rectangle is 6.0 m$^2$. The two fractional uncertainties are $\dfrac{0.1}{2.0} = 0.05$ and $\dfrac{0.3}{3.0} = 0.10$.

The sum of these uncertainties is 0.15, which is the fractional uncertainty of the answer. So the absolute uncertainty in the area $= 0.15 \times 6 = 0.90$.

The answer should be expressed as $(6.0 \pm 0.9)$ m$^2$.

Algebraically: when $y = \dfrac{ab}{c}$ then $\dfrac{\Delta y}{y} = \dfrac{\Delta a}{a} + \dfrac{\Delta b}{b} + \dfrac{\Delta c}{c}$

Again, even when division is used to find the answer, the fractional uncertainty is still added.

**When raising a quantity to a power $n$ the fractional uncertainty is multiplied by $n$.**

This follows from the previous result. When $y = a^2$, this is multiplication of $a \times a$, so, using the algebraic rule: $\dfrac{\Delta y}{y} = \dfrac{\Delta a}{a} + \dfrac{\Delta a}{a} = \dfrac{2\Delta a}{a}$

For example, the radius of a sphere is $(10 \pm 1)$ cm. What is the volume of the sphere?

Volume of sphere $= \dfrac{4}{3}\pi r^3$ where $r$ is the radius.

Fractional uncertainty of radius $= \dfrac{1}{10} = 0.1$ so fractional uncertainty of radius cubed is $3 \times 0.1 = 0.3$.

Volume of sphere $= 4190$ cm$^3$ and the absolute uncertainty is 1260 cm$^3$.

The value for the volume of the sphere is $(4.2 \pm 1.3) \times 10^3$ cm$^3$.

**DP ready** **Nature of science**

## A "thought experiment" on how length and area change resistance

Sometimes in science, practical experiments are not needed. You can think about a problem—though you may choose to verify your thinking with an experiment later. Now that you have an understanding of the way that resistors are combined in series and parallel, you can revisit how resistance depends on shape:

**Length:** Imagine two resistors, identical in size and shape, each of resistance $R$, connected in series by a short wire (the wire is assumed to have no resistance). Using the combination rules, the total resistance is $2R$. Now imagine that the connecting wire becomes shorter and shorter—this has no effect on the resistance—until it eventually disappears. The two resistors are now joined together and are twice as long as each original resistor. This indicates that the resistance of a conductor is directly proportional to its length $l$; in symbols $R \propto l$.

**Area:** Imagine the two identical resistors again, but this time connected in parallel. Their combined resistance in parallel is $\frac{R}{2}$. Now imagine that they merge along their length to form a new resistor of the same original length but double the area $A$. So the resistance of this larger resistor has halved when the area doubled. This suggests that $R \propto \frac{1}{A}$.

The results for length and area together give $R \propto \frac{l}{A}$ or $R = k \times \frac{l}{A}$

where $k$ is a constant. It turns out that $k$ is an important constant that relates only to the material of the conductor. You will learn about this shape-independent constant in your Physics Diploma Programme studies.

As an interesting experiment you could use conducting putty, a material that can be moulded into any shape, to test these two predictions about $l$ and $A$.

**DP link**

You will learn more about the treatment of errors and uncertainties in
**1.2 Uncertainties and errors**.

---

### Worked example: Resistors in combination

**WE**

4.  Four resistors are connected as shown.

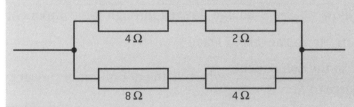

Calculate the total resistance of the network.

*Solution*

The top series pair has a resistance of $4 + 2 = 6\ \Omega$. The bottom series pair has a resistance of $8 + 4 = 12\ \Omega$.

The resistance $R$ of the two parallel pairs is given by

$$\frac{1}{R} = \frac{1}{6} + \frac{1}{12} = \frac{2+1}{12} = \frac{3}{12} \text{ so } R = \frac{12}{3} = 4.0 \quad .$$

**5.** Two resistors of resistance $R_A$ and $R_B$ are connected in parallel. Determine the total resistance of the two resistors.

*Solution*

$$\frac{1}{R} = \frac{1}{R_A} + \frac{1}{R_B} = \frac{R_B + R_A}{R_A R_B} \text{ so } R = \frac{R_A R_B}{R_B + R_A}$$

## Question

6   Calculate the total resistance of each of the following resistor combinations. Sketch the combination as part of your answer:

   a)   $5\,\Omega$ and $18\,\Omega$ in series

   b)   $5\,\Omega$ and $18\,\Omega$ in parallel

   c)   $5\,\Omega$, $10\,\Omega$ and $15\,\Omega$ in series

   d)   $5\,\Omega$, $10\,\Omega$ and $15\,\Omega$ in parallel.

7   A $2\,\Omega$, a $4\,\Omega$ and an $8\,\Omega$ resistor are available. Calculate, using all three resistors,

   a)   the largest combined resistance that is possible

   b)   the smallest combined resistance that is possible.

   Draw the arrangements of resistors as part of your answer.

## Electrical energy revisited

The importance of measuring electrical energy transfers in terms of J C$^{-1}$ was discussed earlier in the chapter. But we still need to be able to describe the total energy transfer (in joules) and the power output (in watts) for devices. Can we obtain these from quantities that we know about already?

From the definition of charge, current can be written as

$$\text{current} = \frac{\text{charge flowing through the component}}{\text{time for charge to flow}}.$$

Potential difference is $\dfrac{\text{energy transferred in a component}}{\text{charge flowing through the component}}$.

When these are multiplied together

$$\frac{\cancel{\text{charge flowing through the component}}}{\text{time for charge to flow}} \times \frac{\text{energy transferred in the component}}{\cancel{\text{charge flowing through the component}}}$$

The "charge flowing" term cancels to leave

$\dfrac{\text{energy transferred in the component}}{\text{time for charge to flow}}$, which is the definition of power.

For electrical components:

power dissipated in the device = potential difference across the device × current in the device

or $P = V \times I$

This expression can be extended using the definition of resistance:

$$P = V \times I = (IR) \times I = I^2 R, \text{ and } P = V \times I = V \times \frac{V}{R} = \frac{V^2}{R}.$$

Since power $= \dfrac{\text{energy transferred}}{\text{time taken}}$,

the energy $E$ transferred by a device in a time $t$ is $P \times t = V \times I \times t$.

### Worked example: Resistors in combination

**6.** Two 8 Ω resistors are connected in parallel and then in series to a 2 Ω resistor. The circuit also contains a cell of emf 12 V.

Calculate:

**a)** the combined resistance of the 8 Ω resistors

**b)** the total resistance in the circuit

**c)** the current in the cell

**d)** the power transferred in the 2 Ω resistor

**e)** the pd across one of the 8 Ω resistors.

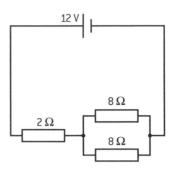

*Solution*

**a)** The combined resistance of the 8 Ω resistors is $\dfrac{1}{R} = \dfrac{1}{8} + \dfrac{1}{8}$
so $R = 4\ \Omega$

**b)** The total resistance (series) $= 4 + 2 = 6\ \Omega$

**c)** The current is $I = \dfrac{V}{R} = \dfrac{12}{6} = 2\ A$

**d)** The power in the 2 Ω resistor $= I^2R = 2^2 \times 2 = 8\ W$

**e)** 4 V is dropped across the 2 Ω ($V = IR$) resistor so $(12 - 4) = 8\ V$ must be dropped across both 8 Ω resistors.

### Question

**8** A cell of emf 3.0 V is connected to a 3.0 V, 3 W lamp in parallel with a 3.0 V, 24 W lamp as shown.

Calculate:

**a)** the current in each light bulb

**b)** the current in the cell

**c)** the power transferred by the cell.

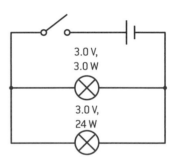

### The potential divider

The **potential divider** is a convenient way to produce a variable potential difference for use in a laboratory circuit. The basic form (figure 11) consists of two resistors, of values $R_1$ and $R_2$, arranged in series with a power supply.

Because it is a series circuit, all the components have the same current. The total pd across the resistors is equal to the emf of the supply. So

$$I = \frac{\text{sum of pd across resistors}}{\text{total resistance}} = \frac{V_{\text{supply}}}{R}$$

We know that $V_1 = IR_1$, $V_2 = IR_2$ and $R = R_1 + R_2$.

With some rearrangement (try this yourself!):

$$V_1 = \frac{R_1}{(R_1 + R_2)}V_{\text{supply}} \text{ and } V_2 = \frac{R_2}{(R_1 + R_2)}V_{\text{supply}}.$$

As an example, if the power supply has an emf of 6.0 V and $R_1$ is 10 Ω with $R_2$ as 30 Ω, then $V_1$ is 1.5 V and $V_2$ is 4.5 V. The $V_1 : V_2$ ratio is the same as the ratio $R_1 : R_2$.

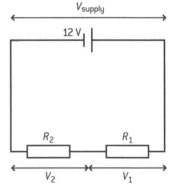

**Figure 11.** Basic potential divider arrangement

### DP link

All practical cells have an internal resistance. This is the reason why cells become hot while charging or discharging.

You will learn about internal resistance in **5.3 Electric cells**.

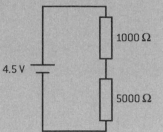

**Question**

9   A potential divider consists of two resistors and a battery as shown.

   **a)**   Calculate the pd across each resistor.

   **b)**   Repeat this calculation when another 5000 Ω resistor is connected in parallel with the first 5000 Ω resistor.

1000 Ω

4.5 V

5000 Ω

## 2.3 Magnetism at work

Magnets and their effects have been studied in many parts of the world from earliest times. There are writings about magnetism in texts from China (6,000 years ago), from Ancient Greece and from India. In 1600, the English scientist William Gilbert published his account of magnetism derived from experiments with a model Earth; he concluded that the Earth acted as a magnet. Two hundred years later, a professor from Copenhagen called Hans Christian Ørsted made the connection between electricity and magnetism that we will explore later in this chapter.

### Permanent magnets

Iron, cobalt and nickel are the only *ferromagnetic* elements. When these metals or their alloys are magnetized and suspended so that they are free to move, they align themselves geographically north–south. The end of the magnet that points towards the geographic north pole is known as the "geographic north-seeking pole", shortened to "north pole" or often just "N". The end pointing south is called the "south pole" or "S". In other words, the Earth's magnetic pole that we call "north" has the opposite polarity to the pole we call "north" on a bar magnet.

---

**Key term**

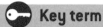

**Ferromagnets** have the following properties.

- Like poles (that is, N and N or S and S) repel; unlike poles (N and S) attract.

- The magnetic force increases as the poles get closer together.

- A magnet can induce magnetism in a nearby unmagnetized ferromagnet.

- Some ferromagnets, typically alloys known as **hard** magnetic materials made from iron and other elements, are permanent magnets and only lose their magnetism under extreme conditions.

- Some, such as pure iron, are magnetic only in a magnetic field and lose the effect when the field is removed; these are known as **soft** magnetic materials.

---

**DP ready**   **Nature of science**

### Other types of magnetism

All magnetic behaviour, including the ferromagnetism that you will study for the IB Diploma Programme, is caused by the spin of electrons around the nuclei of atoms. The other principal forms of magnetism are paramagnetism (where there are unpaired electrons in atomic orbitals) and diamagnetism (a weak effect, present in all materials, which opposes an external magnetic field applied to it).

When ferromagnetic materials are unmagnetized, their electron spins are random. But when a magnetic field is applied the spins line up. In permanent ferromagnets, when the external field is removed, the electron spins do not become random again. This explains why heating and hammering can destroy even permanent magnets, as the thermal effects knock the spins out of alignment.

Another way to destroy a permanent magnet is to put it into a coil that has an alternating current and therefore an alternating field. When the current is reduced to zero, the magnetic field in the magnet will be reduced to zero too.

## Magnetic field lines

We used electric field lines as a way to visualize electric fields. Magnetic field lines are used in the same way; they do not actually exist but they help us to imagine the shape and structure of a magnetic field.

**Internal link**

You met the idea of electric field lines in **2.1 Electric fields and currents**.

- Magnetic field lines show the local direction in which a north pole will move when free to do so (unlike electrical charges, poles always occur in north–south pairs—no free single magnetic pole has been observed). The line direction is therefore from north to south.

- The line density of the field lines gives an indication of the field strength. The closer the field lines are together, the stronger the field.

- Field lines do not cross (if they did, this would imply that the direction is ambiguous, and that can never be the case).

- Field lines are as short as possible; they act in the same way as elastic threads.

Many of the properties of ferromagnets become clear if considered in terms of field lines (figure 12).

Figure 12(a) shows the field due to a single bar magnet. In figure 12(b), the "elastic" field lines between two opposite poles will tend to shorten if possible, meaning that the field will try to move the magnets together. In the case in which two like poles are close together, as in figure 12(c), the absence of lines in the centre indicates a weak or zero field (the poles cancel out giving no force). The implied strong field where the field lines are dense tells us that, if free to do so, the magnets will move apart.

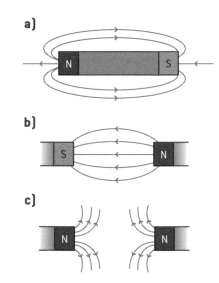

**Figure 12.** Field lines
**(a)** for an isolated bar magnet;
**(b)** between two opposite poles;
**(c)** between like poles close together

---

**DP ready** | **Nature of science**

### Modelling the Earth

In 1600 William Gilbert concluded that the Earth acted as a magnet (figure 13).

Notice the way round the magnet is drawn: the field line directions conform to our labelling of a "north" pole as short for "geographic north-seeking" pole.

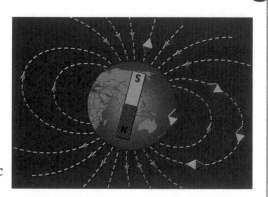

**Figure 13.** The magnetic field of the Earth

The field lines are parallel to the Earth's surface only at the equator. A magnetic compass is of little use at the magnetic poles, where a magnet will be at 90° to the surface.

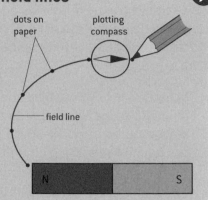

**DP link**

You will learn about the magnetic effects of an electric current in **5.4 Magnetic effects of electric currents**.

## Magnetic effects of a current

Ørsted showed that magnetism is linked to an electric current. It is one of three effects (the other two being heating and chemical effects) that occur when charge flows. Once again, we can imagine the magnetic field using field lines.

A magnetic field is created when charge moves in a straight wire. Figure 14 shows the shape.

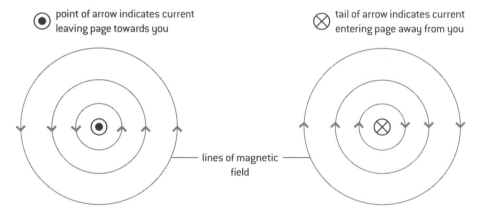

**Figure 14.** Magnetic field shape for a long straight wire

Notice the following points:

- the lines form circles centred on the wire (no poles exist in this arrangement)
- the lines become further apart the greater the distance from the wire (the field strength decreases with distance from the wire)
- the direction of the current is linked to the direction of the field lines.

A direction rule, the **right-hand grip rule**, links the direction of the current and the direction of the field lines (figure 15). It is so called because, when a right hand grips the wire with the thumb pointing in the conventional current direction, then the fingers of the hand curl in the direction of the field.

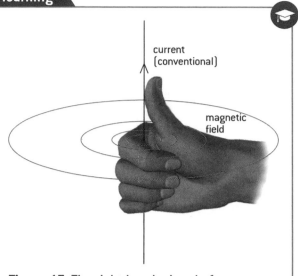

**Figure 15.** The right-hand grip rule for magnetic field in a current-carrying wire

Knowing the field for the straight wire, you can deduce field patterns for other configurations of current. The field due to a solenoid (a long, thin coil) is an example.

- Imagine the long wire being coiled up into the solenoid shape.
- Now think about the field between two adjacent turns on the wire (figure 16(a)). The current is in the same direction in both turns, so the fields must cancel out between the turns and add outside the turns. Near the turn itself the individual field is strong and almost circular.

**a)**

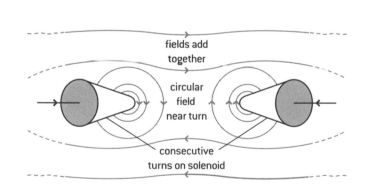

**b)**

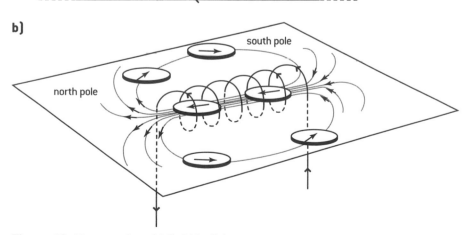

**Figure 16.** How a solenoid field builds up

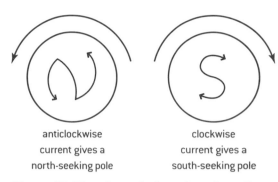

anticlockwise
current gives a
north-seeking pole

clockwise
current gives a
south-seeking pole

**Figure 17.** Direction rule for poles of a coil or solenoid

- The net effect gives the field shown in figure 16: straight in the centre of the solenoid and very similar to that of a bar magnet outside.
- We can predict which end of the solenoid is the north pole, because we know that a bar magnet has field lines coming out of the north pole end. The north pole for the solenoid must be the same. This is the end that, when looking into the solenoid, has the conventional current anticlockwise. There is a direction rule for this too (figure 17).

---

### DP ready  Nature of science

On what does the magnetic field strength of a solenoid depend? We can perform another thought experiment to predict how the magnetic field strength in a solenoid can be changed.

- Increase the current—imagine two sets of wires wound together, each with the original current. This would double the field lines and therefore double the field in the centre.
- Make the turns closer together. This would increase the field lines too, because their density must be larger.
- Put a piece of soft iron in the centre. It will strengthen the total field because a magnetic field will be induced in the iron while the current is on. This field will be in the same direction as the field due to the current and will add to it.

---

### Question

10 Sketch the magnetic field pattern due to a single-turn coil. You can carry out some internet research to answer this question.

11 A solenoid is connected to an electric cell as shown on the left.

a) State and explain **three** ways in which the strength of the field inside the solenoid can be increased.

b) Predict the magnetic pole at the left-hand end of the solenoid.

### The relay

Putting soft iron into the centre of a solenoid or coil leads to the idea of a relay, where a small current produces a large magnetic field in a soft iron core. This can be used to switch a larger current on and off. Figure 18 shows the construction of a relay.

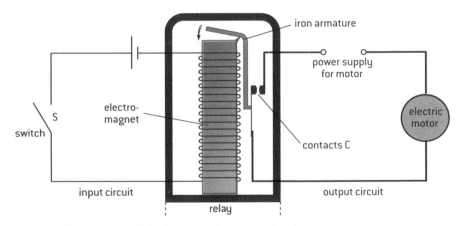

**Figure 18.** A relay and its input and output circuits

The input circuit (small current) is on the left; the output circuit (large current) is on the right. With current in the input circuit the iron core becomes magnetized and attracts the soft iron armature at the top of the solenoid, thus closing the pair of contacts C. There is now current in the right-hand circuit, so the electric motor operates. Switch off the current at S and the two pieces of iron in the circuit become unmagnetized, the contacts C open again because of the spring, and the motor switches off.

## Forces between moving charges

Magnetic field patterns help to explain interactions between wires carrying currents.

**Currents in two straight wires:** Consider two parallel metal foils that carry equal currents (figure 19).

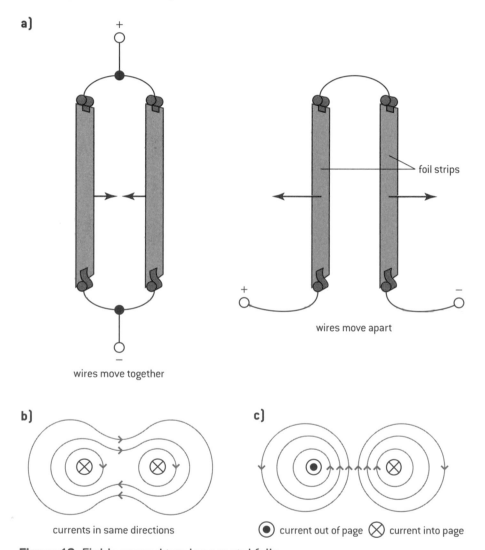

**Figure 19.** Fields around two long metal foils

The currents can be in the same direction (figure 19(a), left) or in opposite direction (figure 19(a), right). The diagrams show the magnetic field patterns and enable us to predict what happens: when the currents are in the same direction the field lines will try to shorten, so the wires move together. When the currents are in opposite directions there is a strong field in the space between the wires which the system will try to prevent. The forces in this case cause the wires to repel and move apart if they are free to do so.

An alternative view shown in figure 19(b) and (c) is that each wire has its own circular magnetic field, inside which the second wire sits. As a result a force acts on the second wire. Equally, the first wire is sitting inside the magnetic field of the second so equal and opposite forces act on both wires.

> **Key term**
>
> The **ampère** is defined so that when there is a current of one ampère in two parallel straight wires of infinite length and of negligible cross-section placed one metre apart in a vacuum, there is a force between them of $2 \times 10^{-7}$ N for every metre of their length.

**A current in a wire in a uniform magnetic field:** One of the most important cases (certainly in terms of its impact on society) is what happens to a current-carrying wire in a uniform magnetic field (figure 20).

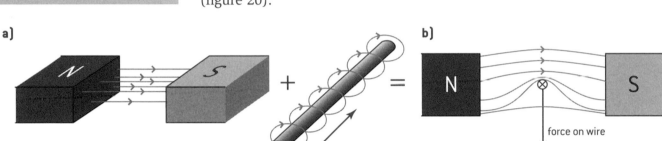

**Figure 20.** A current-carrying wire in a uniform magnetic field

The two magnetic fields have completely different shapes: the field due to the wire is circular and centred on the wire, the uniform magnetic field (shown in figure 20 produced by two poles) has straight, evenly spaced field lines between the magnet poles.

What happens when the circular field is inside the other field? The uniform field is directed from left to right and the current field is clockwise. The two fields interact to give a new field shape with lines more closely packed above the wire than below. The system tries to prevent a high density of field lines like this and can only do so by producing a force that acts downwards on the wire. If the wire is free to move it will be accelerated downwards until out of the uniform field. This effect is sometimes called the **catapult field** because the bent field lines above the wire are like the elastic strings in a catapult.

## The direct current (dc) electric motor

The catapult field described before enables us to transfer energy to a kinetic form in the **dc electric motor**.

Figure 21(a) shows a single-turn coil connected to an electric cell placed in a magnetic field. When there is current in the coil then the magnetic fields interact and distort as before; but this time there are two circular fields and two distortions. The overall result is shown in figure 21(b), and you will see that there must be two equal forces in opposite directions acting on the coil.

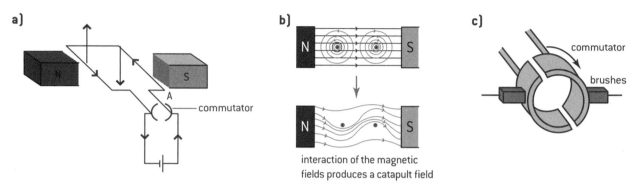

a)

commutator

A

b)

N      S

N      S

interaction of the magnetic
fields produces a catapult field

c)

commutator

brushes

**Figure 21.** The production of the catapult field

These forces rotate the coil. A device called a commutator (figure 21(c)) allows the coil to continue to spin so long as current is present in the coil. The commutator ensures that the left side of the coil always has the direction of current out of the paper and the right side of the coil always has the direction of current into the paper. This is important, as the left side must always be forced upwards and the right side always forced downwards, otherwise the coil will not rotate consistently in the same direction.

What can be done to the coil and the arrangement to increase the forces acting on the coil? The answer is:

- increase the current
- increase the number of turns
- increase the strength of the uniform magnetic field
- increase the coil area, as this will mean a greater length in the field.

## 2.4 Electromagnetic induction

In *2.3 Magnetism at work* we saw that when a wire carrying current is in a magnetic field then a force (and possibly motion) occurs. You could write:

electric current + magnetic field → motion

In fact, the motion of a conductor relative to a stationary magnetic field can also give rise to a current.

motion + magnetic field → electric current

This effect of generating an emf is known as **electromagnetic induction.**

## Practical skills: Electromagnetic effects

Some basic experiments in electromagnetism can be carried out very simply. All you need is a bar magnet and a length of wire wound into a coil. Make sure the magnet can fit into the centre of the coil. Connect the wire to the terminals of a sensitive ammeter and you are ready to go.

- Observe the meter very carefully as you push the magnet quickly into the coil. What happens, and when?
- Try different motions: pull the magnet out, reverse the poles of the magnet.
- What happens when the magnet is moving whilst completely in the coil?
- What happens when the magnet is stationary?

Figure 22 shows some typical results for an analogue ammeter with a needle.

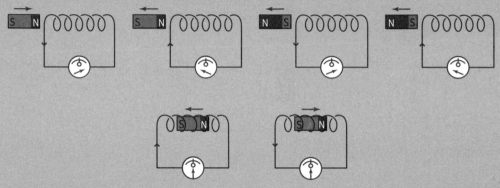

**Figure 22.** Electromagnetic effects

It is also possible to draw conclusions about the direction of the induced current. You may need to carry out a careful experiment to determine the direction of conventional current in the meter to help here. You will require a cell with marked polarity, a large-value resistor (to protect the meter), and connecting leads to do this.

Experiments with electromagnetic induction show that an emf (and therefore a current) is induced in a conductor when:

- the conductor moves relative to the magnetic field
- the magnetic field moves relative to the conductor
- the magnetic field strength changes but the conductor and magnetic field do not move relative to each other.

The magnitude of the emf is increased by making the change (whichever one of the three) more quickly or by increasing the size of the magnetic field.

The final, very important, conclusion is that the induced emf is in a direction that always opposes the change. This is always true, whichever of the three methods is used to produce the emf. A thought experiment tells you why: if the emf were to reinforce the change, then the change would be enhanced and would happen even more quickly; we would obtain energy from the system for nothing! This is never possible (it would permit perpetual motion machines) and so the induced emf and its effects must always oppose the action that is inducing them.

The consequences for the basic experiment are easily seen (figure 22).

When the north pole of a magnet is moved into a coil then the charge flow in the coil must lead to a north pole at the end nearest the magnet, as this way the system can resist the movement of the magnet (by repelling north pole with a north pole).

 **Internal link**

Notice that the term emf is used here rather than pd, because the energy is transferred from a mechanical form into an electrical form (just as a cell transfers from chemical to electrical). The distinction is covered at the end of **2.1 Electric fields and currents**.

When the north pole is moved away from the coil then a south pole appears at the end of the coil next to the magnet because, by attracting the magnet, the coil will tend to reduce the speed at which it moves.

The rule that governs this behaviour, *Lenz's law*, was stated by a Russian scientist, Emil Khristianovich Lenz.

**DP ready    Approaches to learning**

There is a direction rule to relate the induced current direction and the direction of the magnetic field to the motion of the conductor. A few moment's thought will convince you that this must be the opposite of Fleming's left-hand rule to take account of Lenz's law. For electromagnetic induction use **Fleming's right-hand rule**. Figure 23 gives you both rules: remember that the left-hand rule is for motor effects, the right-hand rule for induction effects.

Learn whichever way is best for you.

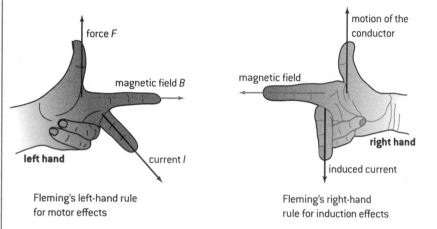

Fleming's left-hand rule for motor effects

Fleming's right-hand rule for induction effects

**Figure 23.** Fleming's direction rules: left hand for motor effects, right hand for induction effects

**DP link**

You will study electromagnetic induction in **11.1 Electromagnetic induction**.

## The alternating current (ac) dynamo

We saw how a motor can be made to rotate when a direct current is in the coil of the motor. The rule motion + magnetic field → electric current seems to suggest that if we move (rotate) a coil in a magnetic field then it should produce an electric current. And so it does.

Figure 24 shows a coil rotating inside a magnetic field. The way in which the current is taken from the generator is, however, different from that of the commutator in the dc motor. Here each brush is permanently connected to one side of the coil (remember, the brushes do not move but the circular collars, known as slip rings, do). The graph (c) shows how the output current varies with time for one cycle, the typical output of an **alternating current**, first positive and then negative. A **direct current** has a steady value that does not vary with time. The output will be a maximum when the coil is horizontal in figure 24 and zero when the coil is vertical at 90° to the magnetic field direction.

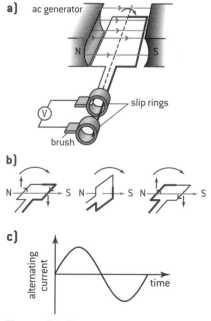

**Figure 24.** The ac generator

12 State **three** ways in which the output voltage from an ac generator can be increased.

13 Identify **two** changes that occur to the current output of an ac generator when the frequency of the coil rotation is doubled.

### The transformer

A transformer is a device that can change the potential difference of an ac supply from one value to another. Transformers come in many sizes and shapes, from the very large ones used in power generation stations to tiny transformers used in electronic circuits.

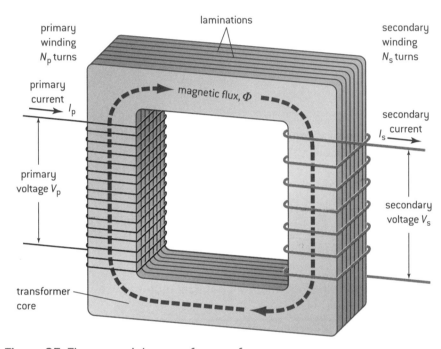

**Figure 25.** The essential parts of a transformer

The main components of a transformer (figure 25) are:

- a primary (input) coil
- a core that is made of soft iron, and
- a secondary (output) coil.

The transformer works in the following way:

- There is an alternating current in the primary coil, so the charge moves first in one direction and then in the other on a regular cycle; there are normally 50 or 60 of these cycles every second for domestic power supplies.

- Because of this current, a magnetic field is set up in the centre of the primary coil. This field alternates (reverses) like the current. The coil is wound on the transformer core, which is made of soft iron and can be easily magnetized and demagnetized, allowing the magnetic field lines to go in a complete circle around it. This circulation of field is known as a **magnetic flux**.

**DP link**

You will learn about the distinction between magnetic flux and magnetic field when you study **11 Electromagnetic induction**.

- As the flux forms a circle, it must also go through the centre of the secondary coil. Therefore, this secondary coil also has a constantly changing flux (magnetic field) inside it.
- This is the condition required for an induced emf to be produced across the coil, and this emf leads to an induced current in the coil when there is a resistor connected across the output.

Analysis shows that the ratio of the primary voltage $V_p$ to the secondary voltage $V_s$ is equal to the ratio of number of turns on the primary $N_p$ to number of turns on the secondary $N_s$: $\frac{V_p}{V_s} = \frac{N_p}{N_s}$. Many transformers are very efficient, close to 100%. Assuming complete efficiency, the power input to the transformer primary will equal the power output at the secondary. In other words: $I_p V_p = I_s V_s$.

For transformers where $N_p > N_s$ (more primary turns than secondary), the primary voltage is greater than the secondary voltage and the transformer is said to **step down** the voltage.

For transformers where $N_s > N_p$, the transformer is said to **step up** the voltage.

## High-voltage transmission

The efficient transmission of energy across a country or even between countries on the same continent relies on transformers to convert low to high voltages (step-up) and vice versa (step-down). When the pd in the secondary is increased by stepping up the voltage, you can see from the equation $I_p V_p = I_s V_s$ that the current drops by the same factor. Transmission at very high voltages is an advantage. Worked example 7 shows why.

**Internal link**

You can review the concept of efficiency in **1.3 Work and energy**.

**DP link**

You will learn about the transformer in more detail in **11.2 Power generation and transmission**, where you will go into detail about the assumptions made in transformer theory and also learn about the losses that occur in real transformers.

---

**Worked example: High and low voltage transmission**

7. Suppose 1 MW of power is transmitted through a transmission cable (another name for a connecting lead) of resistance 0.10 Ω. Calculate the power loss for two cases:
   a) transmission at high current, low voltage: pd 250 V
   b) transmission at high voltage, low current: pd 250 kV (1000 × the first case).

*Solution*

a) For pd 250 V, the current in the cable is $I = \frac{P}{V} = \frac{10^6}{250} = 4$ kA.

This large current leads to a power loss in the cable of

$P = I^2R = (4 \times 10^3)^2 \times 0.10 = 1.6 \times 10^6$ W.

So more energy per second is required to send the current in the cable than is transmitted to the far end! It is, of course, transferred into thermal energy.

b) For pd 250 kV, the current is now $I = \frac{P}{V} = \frac{10^6}{2.5 \times 10^5} = 4$ A, 1000 × less.

The power required in the cable for transmission is now
$P = I^2R = (4)^2 \times 0.10 = 1.6$ W—a considerable saving in energy for the transmission.

A grid system is used in many countries to transmit and integrate power supplies into the national fabric. Figure 26 shows how such a grid can be arranged.

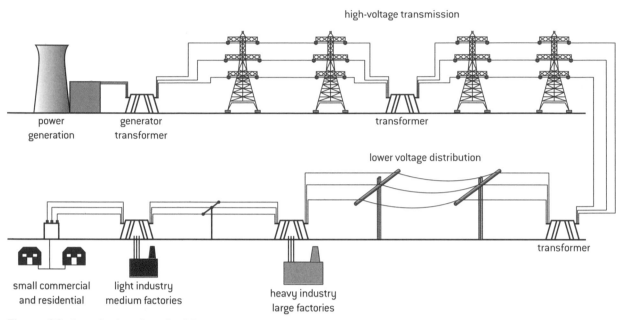

**Figure 26.** A typical national grid system

## Worked example: Calculations involving transformers

8. A power generation station has an output of 25 kV. This is increased to 400 kV using a transformer that has a primary coil with 5000 turns.

   **a)** Identify the type of transformer required.

   **b)** Calculate the number of turns on the secondary coil of the transformer.

*Solution*

**a)** A step-up transformer is needed.

**b)** $\dfrac{V_p}{V_s} = \dfrac{N_p}{N_s}$ so $\dfrac{25\,000}{400\,000} = \dfrac{5000}{N_s}$

$N_s = \dfrac{5 \times 400\,000}{25} = 80\,000$ turns.

## Question

14 A small computer requires a power supply of 12 V, 0.3 A that is supplied by a step-down transformer connected to the 240 V mains supply. The primary coil has 2300 turns. Assume that the transformer is 100% efficient.

   **a)** Calculate the number of turns on the secondary coil.

   **b)** Calculate the current in the primary coil.

15 A 100% efficient transformer has a 20-turn primary and a 1000-turn secondary coil.

   **a)** State and explain whether the transformer is of step-up or step-down type.

   **b)** The transformer primary is connected to the mains supply of 240 V. Calculate the output voltage across the secondary coil.

   **c)** Identify where in a national grid system this transformer is likely to be placed.

## 2.5 Practical aspects of electrical physics

Here are some pointers to help you understand electrical circuits and the concepts of charge, current, pd, emf and energy.

- Components must be arranged in complete electrical circuits for there to be a current and to enable energy transfers to take place.

- Components can be in series or parallel arrangements.

- These circuits use simple combinations of electric cells, switches and resistors. Cells provide the energy source, the switch allows charge to flow in the completed circuit or not, and a resistor is a device that transfers electrical energy into thermal energy.

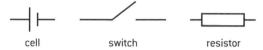

cell         switch         resistor

We always assume that the connecting wires (or leads) in a circuit do not transfer energy from the cell (we say that they have **zero** electrical resistance). This is usually a reasonable assumption in practical work. It is not usual to state this in an examination; it is assumed that the connection links have no effect on the circuit unless you are told otherwise.

| Series circuit | Parallel circuit |
|---|---|
| 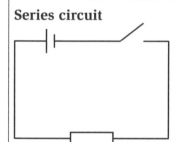 | 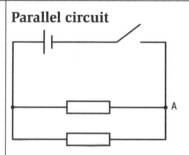 |
| This simple circuit contains a cell, a resistor and a switch. | This circuit contains a cell, two resistors in parallel and a switch. |
| The cell transfers energy to the connecting wires, which carry the energy through to the resistor. In the resistor, energy is transferred from the electrons to the material of the resistor in an internal form (thermal in this case). | The cell transfers energy to the connecting wires. At point A—called a junction—some charge must flow through the top resistor and some must flow through the bottom resistor. The total charge flowing through the cell is equal to the sum of these two charges during any given time interval. The current in the cell must equal the sum of the currents in the two resistors. |
| The energy per coulomb transferred from the cell is transferred to the resistor. The amounts are equal. **The total energy transferred into a circuit per coulomb is equal to the total energy transferred out of the circuit per coulomb.** This is **Kirchhoff's Second Law**. This law is a statement of conservation of energy. | **The current in the wire leading to a junction is equal to the sum of the currents in the wires after the junction.** |
| The charge flowing is the same everywhere in a series circuit. **The current is the same everywhere in a series circuit.** | This law is known as **Kirchhoff's First Law**. This law is a statement of conservation of charge (current). |

**DP link**

You will learn about Kirchhoff's laws and series and parallel resistance arrangements when you study **5.2 Heating effect of electric currents**.

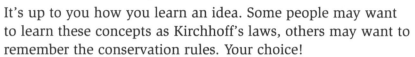

**DP ready** **Approaches to learning**

### Kirchhoff's laws or equations?

It's up to you how you learn an idea. Some people may want to learn these concepts as Kirchhoff's laws, others may want to remember the conservation rules. Your choice!

### Practical skills: Circuits, symbols and diagrams

The translation of circuit symbols into the electrical devices used in the laboratory is an essential skill. You should know all the symbols in figure 27.

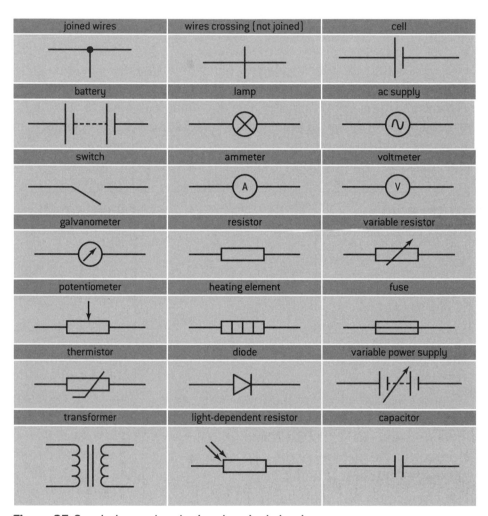

**Figure 27.** Symbols used to depict electrical circuits

### Using meters

Measurements of pd and current are carried out with digital or analogue voltmeters and ammeters. Students sometimes find it hard to recall where these should be placed in a circuit. One way to recall the position of an ammeter is to remember that it measures the charge flowing every second through a component. This means that **ammeters are in series** with the component or the part of the circuit they are measuring.

Voltmeters indicate the energy transferred within a component. So one way to remember the voltmeter's position is to imagine that it needs to compare the energy of the electrons that emerge from a component with the energy of electrons going in—this means **voltmeters are in parallel** with the component they measure. (In fact, voltmeters contain a resistor of accurately known resistance and work by measuring the current in this resistor, which is converted to a pd with $V = IR$.)

### Constructing a working circuit from a circuit diagram

We are going to construct a relatively difficult circuit, a potential divider circuit (introduced in *2.2 Electrical resistance*), which is used to provide a wide range of available pds for a circuit. The circuit diagram is shown in figure 28. The circuit is designed to vary the current in the lamp and to measure the current in the lamp and the pd across the lamp and ammeter.

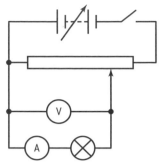

**Figure 28.** A potential divider circuit

1. Begin by laying out all the circuit components on the bench in the same position as in the diagram.

2. Notice that this circuit (figure 28) divides into three parts: the top loop—a cell, a switch and a variable resistor, a middle section—just the voltmeter, and a bottom loop—the ammeter and a lamp.

3. Connect the top loop using connecting leads, but don't switch on yet (figure 29).

4. Now connect the ammeter and lamp together and join them to one end of the variable resistor and its slider. (It doesn't matter whether the loop is geometrically in the middle or at the bottom; they are in parallel.)

5. Finally connect the voltmeter in parallel with the ammeter–lamp loop.

6. Check the circuit against the diagram. Your teacher may wish to check the circuit too. Then set the variable resistor slider (B) to 0 V and switch on.

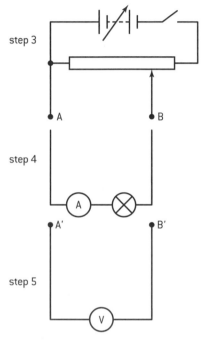

**Figure 29.** Steps in setting up the potential divider circuit

### Rules for drawing circuits

Circuits are always shown as complete. If you want to show a circuit without charge flowing, use a switch in the diagram.

When there are junctions with two or more wires it is usual to put a dot or blob at the intersection. This means that when you see two wires crossing without a dot, these wires do not connect.

---

### Chapter summary

Make sure that you have a working knowledge of the following concepts and definitions:

☐ Like charges repel and unlike charges attract, and the migration of charge can cause an uncharged object to be attracted to a charged object.

☐ Charge is a conserved quantity.

☐ An electric field surrounds a charged object; this electric field can be represented by electric field lines.

☐ When an electric field acts on a conductor then the charge carriers in the conductor are accelerated by the field and can transfer energy to the fixed atoms in the conductor.

☐ The charge carriers in metallic conductors are free electrons released when the atoms bond together to form the solid.

☐ Insulators have few charge carriers; semiconductors can have charge carriers of either sign of charge.

☐ The flow of charge in a conductor is called an electric current. The electric current is the rate at which charge flows past a point in the conductor.

☐ Electric current is measured in ampères; the ampère is a fundamental (base) SI unit.

- ☐ Potential difference and emf are measured in terms of the amount of energy transferred in a device when one coulomb of charge flows through the device.
- ☐ Potential difference (pd) is the term used when energy is transferred from the electrical store to an energy sink that is not electrical; electromotive force (emf) is the term used when energy is transferred from a non-electrical source (typically an electric cell or battery) to an electrical sink.
- ☐ There are rules for setting up circuits and for drawing electrical circuits.
- ☐ Kirchhoff's two laws for electrical junctions and circuits are examples of the conservation of charge and the conservation of energy.
- ☐ Resistance is defined as the ratio $\dfrac{\text{pd}}{\text{current}}$.
- ☐ The V–I or I–V characteristic can be used to determine whether a material is an ohmic conductor (one for which the characteristic is a straight line through the origin when the temperature of the conductor is constant).
- ☐ The resistance of a conductor is directly proportional to the length of the conductor and inversely proportional to the area of the conductor.
- ☐ Resistors can be joined in series or parallel arrangements or in combinations. When in series, the resistance values add to give the total resistance of the combination. When in parallel, the reciprocal of the total resistance is equal to the sum of the reciprocals of the separate resistors.
- ☐ A colour code is used to identify the resistance values of commercial resistors.
- ☐ When a device has a current $I$ with a pd $V$ across it, the energy transferred by the device in time $t$ is $VIt$.
- ☐ The electrical power transferred by the device is $VI$; this can also be expressed as $I^2R$ and $\dfrac{V^2}{R}$.
- ☐ The potential divider arrangement has advantages over a variable resistor arrangement for the supply of a wide range of pd values across a component.
- ☐ Ferromagnetism is a property of iron, cobalt, nickel and their alloys. Ferromagnets will align to a north–south orientation in the Earth's magnetic field.
- ☐ Some ferromagnets are permanently magnetized and are called hard magnetic materials. Such materials can be heated or beaten to destroy their magnetism. Some ferromagnets are only magnetized when in a magnetic field and are called soft magnetic materials.
- ☐ A magnetic field can be described by magnetic field lines that indicate the strength of the field (through their relative density) and the direction of the field.
- ☐ Magnetic fields can be visualized using small compass needles or iron filings.
- ☐ Magnetic effects arise from the flow of charge. A long straight current-carrying wire has a magnetic field that is circular and centred on the wire. A current-carrying solenoid or coil has a magnetic field that resembles that of a bar magnet.
- ☐ The interaction of a uniform magnetic field and the magnetic field of a wire is important as it can lead to the transfer of energy from an electrical to a kinetic form. This is the basis of the relay and the dc electric motor.
- ☐ When a conductor moves relative to a magnetic field, an emf can be induced in the conductor. This effect is the basis of the ac electrical generator.
- ☐ The direction of the emf and therefore any induced current in the conductor is such as to oppose the relative motion of the conductor and magnetic field.
- ☐ Changing the strength of a magnetic field while keeping the conductor stationary can also cause an emf to be induced. This effect is the basis of the transformer.
- ☐ Transformers can step up (increase) the output voltage over the input or step down (reduce) the output voltage relative to the input. For a 100% efficient transformer, the power input to the primary coil is equal to the power output from the secondary coil.
- ☐ Transformers are used to change the value of the alternating pd in a national grid system for the efficient transmission of electrical energy.

## Additional questions

1. In the following diagrams each resistor has a resistance of 5.0 $\Omega$ and the battery has an emf of 12 V.

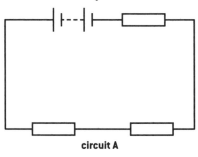

circuit A

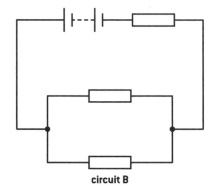

circuit B

    **a)** Calculate, for circuit **A**:

        **i)** the current in each resistor

        **ii)** the power dissipated in each resistor.

    **b)** Calculate, for circuit **B**, the current in each resistor.

    **c)** Discuss the energy transferred in the parallel resistors in circuit **B** compared with that in the series resistors in circuit **A**.

2. The resistance of a metal wire was determined at six temperatures. The results are given in the table.

    **a)** Plot a graph to show the variation of resistance with temperature for the wire.

    **b)** Use the graph to determine the value of resistance $R_0$ at 0°C.

    **c)** Determine the gradient of the graph.

    **d)** The gradient of the graph is equal to $kR_0$ where $k$ is a constant.

        **i)** Calculate, using your answer to **b)**, the value of $k$.

        **ii)** State the unit of $k$.

| Temperature (°C) | Resistance ($\Omega$) |
|---|---|
| 10 | 2.53 |
| 20 | 2.70 |
| 35 | 2.94 |
| 50 | 3.17 |
| 65 | 3.39 |
| 80 | 3.64 |

3. A heater for the rear window of a motor vehicle has six identical heating wires connected as shown on the right. The resistance of each wire is 6.0 $\Omega$. The heater is connected to the 12 V battery of the vehicle.

    **a)** Calculate the current in each wire.

    **b)** An alternative design has all six wires in series. Calculate the current in each wire.

    **c)** Use your answers to **a)** and **b)** to suggest why the parallel arrangement is more effective than the series arrangement.

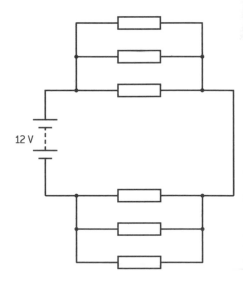

4. A wire carries an electric current between the poles of a magnet.

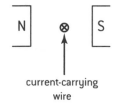

current-carrying wire

    **a)** Draw the magnetic field lines that result from this arrangement.

    **b)** Use your diagram to explain why there is a force acting on the wire due to its interaction with the field of the magnet.

5. Two magnets are placed on a top-pan balance. A long straight horizontal wire passes between the magnets so that a horizontal magnetic field acts at 90° to the direction of the wire. When there is no current in the wire the reading on the balance is
351.20 g. When there is a current in the wire the reading on the balance changes to 352.6 g.

   a) Explain the change in balance reading when there is a current in the wire.

   b) Deduce the direction of current in the wire.

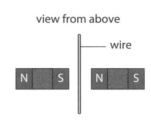

view from above

6. The graph shows the variation of pd $V$ across resistor A with current $I$ in the resistor.

   a) Calculate the resistance of A when the current in it is 0.18 A.

   b) Discuss whether A obeys Ohm's law.

   c) Another resistor B has the same resistance as A at a current of 0.18 A. B obeys Ohm's law. Copy the graph and sketch the variation of $V$ with $I$ for resistor B.

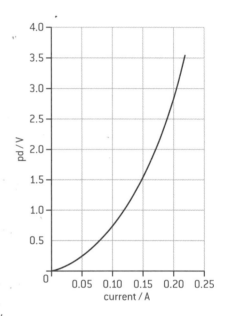

7. Some of the values in the circuit below are missing (grey boxes).

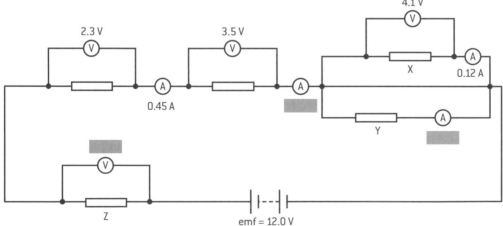

   a) Deduce the values that are missing.

   b) Calculate the resistance of: (i) X , (ii) Y, and (iii) Z.

   c) The battery can transfer a total of 0.85 MJ. Calculate the maximum time for which the battery can operate.

8. Two 12 V, 18 W lamps are connected in parallel with a 12 V, 4.0 W lamp.

   Calculate the total resistance of the three lamps connected in parallel at their normal working temperature.

# Thermal physics

> ❝ No fact in the world is instant, infinitesimal and ultimate, a single mark. There are, I hold, no atomic facts. In the language of science, every fact is a field—a crisscross of implications, those that lead to it and those that lead from it. … We condense the laws around concepts. Science takes its coherence, its intellectual and imaginative strength together, from the concepts at which its laws cross, like knots in a mesh. ❞
>
> **Jacob Bronowski, *The Sense of Human Dignity* (1956)**

## Chapter context

Thermal physics is a huge subject, looking at energy stores and transfers in all states of matter. In this chapter you will look at **phase changes, temperature** and **internal energy**, and the behaviour of gases, amongst other topics.

## Learning objectives

In this chapter you will learn about:

→ the **states of matter**

→ the meaning of **"temperature"**

→ **internal energy**

→ **pressure**

→ the **gas laws** and **kinetic theory**

→ changes of state and **specific latent heat**

→ changes of temperature and **specific heat capacity**.

 **Key terms introduced**

→ states (or phases) of matter

→ fusion and evaporation

→ temperature

→ internal energy

→ heat

→ absolute zero

→ pressure

→ density

→ specific latent heat

→ specific heat capacity

## 3.1 States of matter

This section deals with the similarities and differences between three states of matter: solids, liquids and gases. We will look at the properties of these phases at the molecular level and at some of the quantities used to describe them. We will also consider models that scientists construct to explain the behaviour of the various phases and behaviours of materials. These models can describe matter at the macroscopic or the microscopic scale.

In some ways this chapter can be regarded as a case study that uses many of the concepts from *1 Motion and force*. Ideas of force, momentum and impulse are used to model the behaviour of matter.

**DP link**

You will learn about the states of matter when you study **3.1 Thermal concepts** in the IB Physics Diploma Programme.

---

**DP ready** **Nature of science**

**Macroscopic and microscopic**

The explanations in this chapter move between a macroscopic, or large-scale, view of the world and a microscopic view. When in macroscopic mode we take the view that objects are continuous pieces of matter without any fine structure. When looking at the world through a microscopic lens we look in detail at the interactions of the individual *particles*.

As an example, in *2 Electric charge at work*, conductors were described as containing charged free electrons (a microscopic view) whose interactions lead to resistance (a macroscopic concept).

As you work through this chapter, try to spot which view is in use at different points.

## Changing state

At the temperatures and pressures of everyday life, materials exist as either **solids**, **liquids** or **gases**. These are called the *states (or phases) of matter*. (There is a fourth, very high-temperature state called a plasma, which we shall mention briefly later in the context of nuclear fusion.) All three states are made up of the same constituents: *atoms*, *ions* or *molecules*.

At all temperatures except the very lowest, all particles move because they have thermal energy. Atomic and molecular motion increases as temperature increases. The amount by which the particles in a material move can be thought of as the feature that distinguishes between the three phases (figure 1).

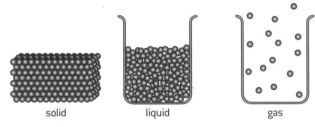

solid        liquid        gas

**Figure 1.** The three principal states of matter

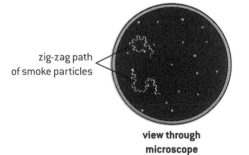

zig-zag path of smoke particles

**view through microscope**

- Solids have shapes that are fixed unless large distorting forces are applied to them. When such a force is applied, the shape will change but the volume will remain almost constant. The particles are strongly bonded and do not change places relative to each other; they vibrate about a fixed position.

- Liquids have particle arrangements that are still bound to each other and this means that liquids have a fixed volume. However, unlike in a solid, in a liquid the particles can slide past each other, allowing the liquid to flow. There is short-range, but no long-range, order.

- Gases have neither a shape nor a fixed volume. They completely fill any closed container in which they are placed. The molecules move at high speed, colliding with the walls of the container and with each other.

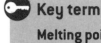

**A brief history of temperature**

Although scientists before the 17th century tried to define temperature and to construct instruments to measure it, the first modern successful attempts date from the 1600s. Galileo and others tried to measure temperature using liquid in round-bottomed flasks from which long tubes emerged, but were largely unsuccessful. A breakthrough came when Ferdinand II, a nobleman from Tuscany, used a sealed vessel that was not affected by air pressure. Since then there have been many developments in thermometry and temperature measurement. The work of Fahrenheit, Romer, Celsius and Kelvin has brought us to our present position, where we understand the relationship between average particle speed and temperature.

## Change of phase

A material can easily change from one phase to another without alteration in its chemical properties. When energy is supplied to ice, it first melts to the liquid form and then boils to produce steam, a gas. These transitions occur at definite temperatures, the *melting point* and *boiling point* (or *freezing point* and *condensing point* when energy is being removed and the phase changes go in the other direction). A later section of this chapter will examine the quantitative changes in energy for these processes.

However, change of phase can occur at intermediate temperatures too. A liquid can *evaporate* to the gas phase below its boiling point (figure 2): this happens when particles near the surface of the liquid and moving at a higher speed than usual have enough energy to leave the surface and become free. This evaporation process reduces the average speed of the particles remaining in the liquid, and as a result the temperature of the liquid decreases too. Ways to increase the rate at which evaporation occurs include increasing the temperature of the liquid (and therefore the number of high-speed particles) and increasing the surface area of the liquid (so that more of the liquid is closer to the surface).

> **Key term**
>
> **Melting point** is the temperature at which a pure substance changes from a solid to a liquid (it melts, or fuses); **freezing** is the change from liquid to solid. **Boiling point** is the temperature at which a pure liquid turns into a gas in bulk (when its vapour pressure is equal to atmospheric pressure). **Condensation** is the reverse process. **Evaporation** occurs at temperatures below the boiling point, when individual particles have sufficient energy to escape from the surface of a liquid.

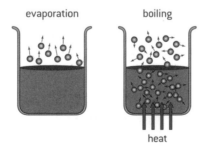

**Figure 2.** Evaporation and boiling

A forced draught across the surface will remove high-speed particles more quickly, thus increasing the rate of loss of particles. You will be familiar with this effect: when you blow across your hand, the liquid particles near the surface of your skin evaporate and the surface cools. The human body uses evaporation to cool down. Sweating occurs when the body temperature rises about 0.5 K above normal. Water is emitted from the pores on the skin and evaporates, transferring energy from the body as it does so.

Boiling can be regarded as a type of evaporation. It occurs when the vapour pressure of the liquid (the pressure it exerts because of the effects of evaporation) is equal to atmospheric pressure. At this point, small bubbles of vapour in the bulk of the liquid begin to expand and move to the surface, releasing the vapour explosively in the effect we call boiling.

### Temperature and thermometers

Whatever phase a material is in, the atoms and molecules in it are always moving. We can define an average speed for these particles. The higher the average speed, the greater the temperature of the object.

### Maths skills: Averages

One of the maths skills you need for DP Physics is that of calculating an *arithmetic mean*. This is also called an *average*. The best way to show how to work out an average is by an example:

Five molecules in a gas have the following speeds: 100 m s$^{-1}$, 200 m s$^{-1}$, 300 m s$^{-1}$, 400 m s$^{-1}$ and 500 m s$^{-1}$.

The average speed is equal to the

$$\frac{\text{sum of all five speeds}}{\text{number of molecules measured}} = \frac{100 + 200 + 300 + 400 + 500}{5}$$

$$= \frac{1500}{5} = 300 \text{ m s}^{-1}$$

The symbol for the average value of a quantity is formed by putting a bar over the symbol for the quantity, so $\bar{v}$ means the average speed of a collection of moving objects.

### Key term

**Temperature** is a measure of average kinetic energy per particle. **Internal energy** is a measure of the total energy, both kinetic and potential, in an object. When energy is transferred from a warm to a cooler object it is called **heat**.

**DP ready    Nature of science**

**Temperature, internal energy and heat...**

...are often confused. *Temperature* is a measure of the average kinetic energy per particle. *Internal energy*, on the other hand, is a measure of the total amount of both kinetic and potential energy stored in an object. Touching a piece of red-hot metal will give you a severe burn, but a white-hot spark from a firework will not hurt you: the spark is at a higher temperature than the metal but contains far less internal energy. When a hot object transfers energy to a cold one, the energy transferred is called *heat*.

hot        cold

**Figure 3.** Thermal energy transfer between hot and cold objects

Imagine two metal objects in contact, with one at a higher temperature than the other (figure 3). When the surface of a hot object is touching the surface of a cool one, energy can travel from hot to cool. This energy is transferred as kinetic energy from the hot object to the cold object. In other words, the average kinetic energy per particle is decreasing for the hot object and increasing for the cold object. When the two values are the same, net heat transfer will cease. When there is no net heat transfer between two objects in contact, then they must be at the same temperature. We rely on this when we use a thermometer: we put the thermometer in contact with the object whose temperature we want to measure, wait until it is likely that thermal equilibrium has been established, and then read the thermometer.

This leads us to the idea of temperature scales, ladders of temperature that allow us to compare different things at different temperatures. Early scales were based on two fixed points—defined temperature points—with the temperatures between them divided into a defined number of equal intervals. The Celsius scale has the melting point of ice (0°C) and the boiling point of water (100°C) as its fixed points, at a standard atmospheric pressure of 101 325 Pa. The interval between them is divided into 100 degrees.

An earlier everyday scale was that due to Gabriel Fahrenheit, who produced the first mercury thermometer. There is some doubt about the exact fixed points he chose and, in any event, the scale was later redefined so that the freezing point of water is set to 32°F and the boiling point of water to 212°F, giving a temperature interval between these of 180 degrees.

## DP ready  Nature of science

### Thermometers

Thermometers come in many forms, some of which you will be familiar with already, like the digital medical thermometer and the mercury-in-glass thermometer. The latter relies on the phenomenon of *expansion*.

mercury

bulb    thread

When the temperature of a liquid or a solid increases, it expands (with a few exceptions); in other words, it increases in volume. This is because as the internal energy of the material increases, the particle vibrations increase and so the dimensions of the whole material tend to get larger.

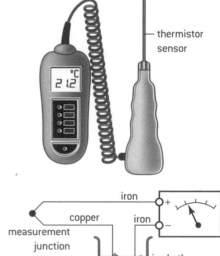

thermistor sensor

Perhaps not so familiar are thermistor-based thermometers, which are based on the resistance properties of a semiconductor substance. Another electrical type, but working in a fundamentally different way, is the thermocouple thermometer, in which two dissimilar metals are joined together, for example iron–copper–iron. When the two junctions are at different temperatures then there is a small emf between them. This can be measured with a sensitive meter and the temperature difference deduced from the emf reading.

iron

copper    iron

measurement junction

reference junction

ice bath (0°C)

thermocouple thermometer

### Maths skills: Equals and equivalent

There is a difference between the symbol =, meaning "equals" and ≡, which means "equivalent".

Equal means "the same"; equivalent means "having the same effect".

The present scientific scale is due to William Thomson, Lord Kelvin, an influential British scientist of the 19th century. He realized that the average kinetic energy per particle would make a good theoretical scale. He defined the lower fixed point to correspond to zero kinetic energy, the temperature at which there is no motion—this is called *absolute zero* for this reason. He then chose the temperature interval to be the same as that of the Celsius scale so that 0 K ≡ −273°C, the freezing point of water is 0°C = 273 K, and the boiling point of water is 373 K = 100°C.

Note that it is not usual to write a degree sign in front of the "K" as the temperature is called a "kelvin" and is an absolute unit, whereas in Celsius it is "degree Celsius" and you must write the "°" sign before the "C".

### Worked example: Converting between degrees Celsius and kelvin

1.  Convert the following temperatures:
    a)  135 K into degrees Celsius;

    b)  375°C into kelvin.

*Solution*

   a)  0 K is defined to be −273°C and one kelvin is the same temperature change as one degree Celsius, so
   135 K = −273 + 135 = −138°C

   b)  0°C is 273 K so 375°C = 375 + 273 = 648 K

### Question

1  Complete the table for the following temperatures (you will have to do some research).

| | Temperature / K | Temperature / °C |
|---|---|---|
| average temperature of Earth | | 16 |
| melting point of zinc | 693 | |
| temperature of the visible surface of the Sun | 5778 | |
| average temperature of the human body | | 37 |
| boiling point of liquid nitrogen | | |
| freezing point of mercury | | |
| surface temperature of dry ice (solid carbon dioxide) when it sublimes from a solid directly to a gas | | |

The second quantity to discuss as we look at the properties of materials is *pressure*. Pressure is associated with all three states of matter but is of considerable importance for a gas.

### Solid pressure

This is a familiar phenomenon. A solid resting on a surface exerts a force and therefore a pressure. The pressure is determined from the weight and the contact area.

> **Worked example: Calculating pressure (1)**
>
>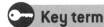
>
> 2. A book of mass 2.0 kg rests on a table. The area of contact between the book and the table is 15 cm × 22 cm. Calculate the pressure of the book on the table.
>
> Force exerted by book = 2.0 × $g$ = 19.62 N    *multiplying by g converts from mass to weight*
>
> Area of contact = 0.15 × 0.22 = 0.033 m²    *notice the conversion to metres*
>
> Pressure = $\dfrac{19.62}{0.033}$ = 590 Pa    *round to 2 sf like the data in the problem*

Pressure can help or hinder us. A sharp knife (thus having a small contact area for the blade) exerts a large pressure from a relatively small force and cuts through materials easily and safely. The small contact area of a drawing pin enables it easily to penetrate a cork board to pin up a notice. A ski or a snow shoe has a much larger contact area than a foot and enables free movement without sinking into the snow. On the other hand, a stiletto heel on a shoe easily damages the surface of a wooden floor. Stiletto heels were once banned in aircraft cabins in the days when the metal of the cabin floor was thin and vulnerable.

### Liquid pressure

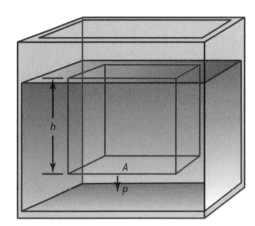

**Figure 4.** Pressure in a liquid

Liquids exert a pressure on their container and on any object in the liquid inside the container. This pressure:

- acts in all directions
- increases with depth
- increases with the density of the liquid.

The last two points are encapsulated in an equation for the pressure $p$ at a depth $h$ in a liquid of *density* $\rho$. Consider a frame with an area $A$ in the liquid (figure 4). The liquid within this frame is in equilibrium with the liquid outside it. The liquid inside the frame has a volume of $A \times h$ and a mass of $\rho \times A \times h$. The weight of the liquid is therefore $g \times \rho A h$ (where $g$ is the acceleration due to gravity).

The weight of this liquid acts downwards on the liquid below it; this is the force in the equation that defines pressure. The pressure $p$ at the depth $h$ in the liquid is therefore

$$p = \frac{\rho A g h}{A} = \rho g h$$

This equation confirms the second and third points above.

> 🔑 **Key term**
>
> **Pressure** is a scalar quantity and is defined to be pressure = $\dfrac{\text{force}}{\text{area}}$.
> For a solid, the force is measured at 90° to the surface.
> The unit of pressure is the pascal (Pa), named for Blaise Pascal, a polymath who lived in the 17th century. The unit is an abbreviation for N m⁻², which can be expressed in fundamental (base) SI units as kg m⁻¹ s⁻².
>
> Other units are possible. The millimetre of mercury (mmHg) is discussed below, and the atmosphere (atm) is sometimes used. The value of standard atmospheric pressure, 1 atm = 101 000 Pa.
>
> Weather forecasters use another non-SI unit of pressure, the bar (b). This is related to the pascal in a simple way: 1 mb ≡ 100 Pa. Thus 1 b = 1000 mb (millibars) is roughly 1 atm or 100 000 Pa.

> 🔑 **Key term**
>
> **Density** $\rho$ is defined as $\rho = \dfrac{\text{mass of an object}}{\text{volume of the object}} = \dfrac{m}{V}$
> The symbol is the Greek letter rho (pronounced "row"). The units of density are kg m⁻³.

## Worked example: Calculating pressure (2)

3. A swimming pool contains water with a density of $10^3$ kg m$^{-3}$. Calculate the pressure due to the water at a depth of 4.5 m. Ignore the effect of the atmosphere.

**Solution**

$p = \rho gh = 1000 \times 9.81 \times 4.5 = 44\,145$ Pa = 44 kPa    (rounded to 2 sf to agree with the data)

(If we include the pressure of the atmosphere pressing on the water surface, then we would need to add another 100 kPa onto this answer.)

4. An elephant has a weight of 25 kN. Each of the elephant's four feet has an area of 0.050 m$^2$. Calculate the pressure that the elephant exerts on the floor.

**Solution**

The total area of the elephant's feet is 0.20 m$^2$. The pressure is therefore $= \dfrac{25000}{0.20} = 1.3 \times 10^5$ Pa.

### The U-tube manometer

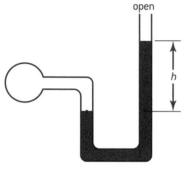

open

**Figure 5.** U-tube manometer

The U-tube manometer is used to measure pressure differences between two gases; typically one of these is air at atmospheric pressure. It operates on the principle that $p = \rho gh$ (figure 5).

The pressure **difference** $\Delta p$ = liquid density $\rho \times g \times$ height difference between the two arms $h$. This instrument is so frequently used that the value for the pressure difference is sometimes quoted, not in pascal, but in length units of the liquid in the manometer. When water is used and the difference in height between liquid levels is 45 mm, then the pressure is quoted as "45 mm of water pressure". Another liquid frequently used in the U-tube manometer is mercury (although safety precautions must be taken to prevent the poisonous mercury vapour from reaching the lungs of a user), hence the unit mmHg.

## Worked example: Calculations involving a U-tube manometer

5. a) Calculate, in Pa, the pressure difference that will cause a U-tube manometer to have a height difference of 3.0 cm of mercury pressure between the arms. The density of mercury is $1.36 \times 10^4$ kg m$^{-3}$.

   b) State this pressure difference as a percentage of atmospheric pressure. Atmospheric pressure is 101 kPa.

**Solution**

   a) The height difference is 0.030 m. So $\Delta p = \rho g \Delta h = 1.36 \times 10^4 \times 9.81 \times 3 \times 10^{-2} = 4.00 \times 10^3$ Pa. This should be stated as 4.0 kPa as the height difference was only specified to 2 significant figures.

   b) $\dfrac{4}{101} \times 100\% = 4\%$

## Maths skills : Calculating percentage

When you are asked to calculate the percentage that one value has compared to another, divide the two quantities and multiply by 100. Express the answer with the % sign after it.

Percentage values and fractional values never have units, because the numerator and denominator have the same units, which therefore cancel.

## The mercury barometer

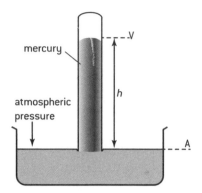

**Figure 6.** Mercury barometer

This can be regarded as a variant of the manometer, in which the U-tube has been straightened out and one end submerged in mercury (figure 6). This means that the mercury level at A is always at atmospheric pressure. The other end of the manometer has been sealed, trapping a vacuum at V above the mercury column. Following the idea of the manometer, the height difference $h$ between the mercury levels is now equivalent to the pressure difference between a vacuum ($\equiv 0$ Pa) and the atmosphere. So, this is the atmospheric pressure expressed as a length $h$ of mercury.

---

### Worked example: Calculations involving a mercury barometer

6. The density of mercury is $1.36 \times 10^4$ kg m⁻³. Calculate the length that a mercury barometer will have when the atmospheric pressure is 101 kPa.

   Rearranging the equation gives:

   $$\Delta h = \frac{p}{\rho g} = \frac{1.01 \times 10^5}{1.36 \times 10^4 \times 9.81} = 0.757 \text{ m or about 760 mm.}$$

---

## 3.2 Gas laws

### Gas pressure

Gases exert pressure on the sides of their containers. How does this force arise?

We saw that the particles that make up a gas are moving independently throughout the volume of the container. They do not all move at the same speed, but with a range of speeds that goes from zero to speeds much greater than the speed of sound. It is possible, therefore, to define an average speed for these particles. In *3.1 States of matter* the average speed of particles was explained to be a measure of the temperature of the gas.

As they move, gas particles collide with each other and with the container wall. It is the microscopic interaction of the particles with the wall that gives rise to the macroscopic quantity we call pressure.

Imagine a single gas particle as it moves up to the wall. Assume that the path of the particle is at 90° to the surface (figure 7).

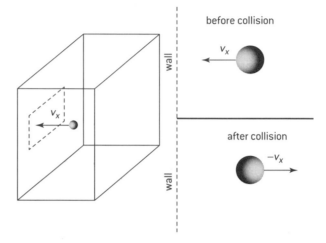

**Figure 7.** Momentum transfer in a gas–wall collision

As the particle approaches the wall at speed $v_x$, atomic forces begin to act between the particle and the atoms in the wall. This force is repulsive (otherwise the particle would attach itself to the wall). The particle reverses its direction and moves away from the wall along the path on which it approached.

What does a study of the mechanics of this situation tell us? It reminds us that momentum is conserved as no external force acts on the particle–wall system. This means (as the system cannot gain momentum) that the velocity of the particle after the collision must be exactly opposite in direction to the original trajectory, that is, $-v_x$; the speed is unchanged but the velocity is changed. So, in algebraic terms, if the original momentum is $+p$, then the final momentum after collision with the wall is $-p$. The change in momentum is therefore $-2p$ $(-p - (+p))$. As momentum is conserved, the wall undergoes a momentum change of $+2p$. We know that this momentum change can be thought of as a force × time taken for the momentum change, and therefore a force acts on the wall to the left in the diagram (from the inside to the outside of the container).

The particle then travels to the opposite wall and repeats the momentum change but this time the change, and therefore the force, is in the opposite direction. In a gas, a very large number of particles all contribute to a total force that acts over the whole area of the wall. The pressure arises from the impulse of many particles in many random directions.

This microscopic model of the gas contains many assumptions and simplifications that you will examine in much more detail in the Physics Diploma Programme. But we need to ensure that the model corresponds, at least in part, to the behaviour of real gases. This is best done experimentally.

## Practical skills: Investigating how gases behave

### When the gas temperature is constant

The pump and the valve shown in the apparatus in figure 8 alter the amount of oil in the vertical tube, and this changes the pressure of the gas trapped in the column. It is important to make the changes slowly so that the temperature of the gas remains constant (think what happens to the air in a bicycle pump when you compress the air). Read off the pressure (from the Bourdon gauge) and the volume (from the length of the gas in the tube).

A plot of gas pressure $p$ against gas volume $V$ will **not** be a straight line, but a graph showing the variation of $p$ with $\frac{1}{V}$ will (figure 9).

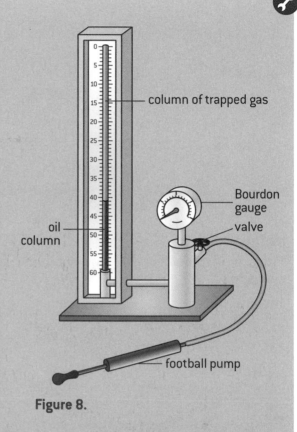

Figure 8.

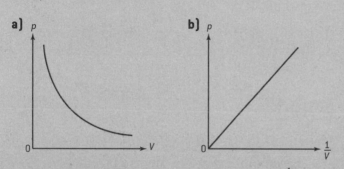

Figure 9. Graph of (a) $p$ against $V$ and (b) $p$ against $\frac{1}{V}$ at constant temperature

The pressure is said to be **inversely proportional** to volume for a gas at constant temperature.

This rule was first suggested by Robert Boyle in the 17th century; it is known as Boyle's law (though you might argue that it is not a law at all).

### When the gas volume is constant

There are many ways to perform this experiment; figure 10 shows one variant.

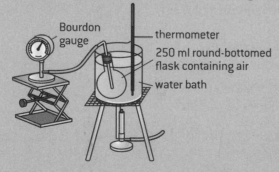

Figure 10. Investigating gas pressure as temperature changes

The Bourdon gauge measures the pressure of a gas in a round-bottomed flask. This flask will expand only a little as the temperature changes and this change is negligible compared with everything else that is changing in this experiment. Make measurements of pressure over as wide a temperature range (in degrees Celsius) as you can. Plot a graph to show the variation of pressure with temperature $T$.

The graph should be a straight line, and this tells us that pressure is **proportional** to temperature (figure 11). If you calculate the temperature intercept for the pressure to be zero, you are estimating absolute zero in degrees Celsius.

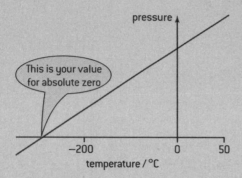

**Figure 11.** A graph of $p$ against $T$ at constant volume

## Maths skills

There are two maths skills required in the gas experiment, proportionality and calculating graph intercepts.

### Direct proportion and inverse proportion

| X | Y |
|---|---|
| 1 | 4 |
| 2 | 8 |
| 4 | 16 |
| 8 | 32 |

| X | Y | X×Y |
|---|---|---|
| 1 | 24 | 24 |
| 2 | 12 | 24 |
| 4 | 6 | 24 |
| 8 | 3 | 24 |

These numbers are in direct proportion. A graph of $X$ against $Y$ would be a straight line going through the origin.

These numbers are in inverse proportion. A graph of $X$ against $Y$ is a curve. A graph of $X$ against $\frac{1}{Y}$ is a straight line. $X \times Y$ is a constant.

### Calculating an intercept on a graph

The problem is calculating the intercept on the $x$-axis of a graph when the intercept does not lie on the graph paper (or, for the gas experiment, anywhere near it!).

One of a number of techniques is to determine the gradient (m) and $y$-intercept (c) of the straight-line graph and then construct the $y = mx + c$ equation of the line.

A substitution of $y = 0$ into the equation will immediately yield the $x$ value at the intercept; so, for figure 12, $0 = 2x + 4$, hence $x = -2$.

Alternatively, you may find that your scientific calculator can compute the intercept when you have keyed in the data point values.

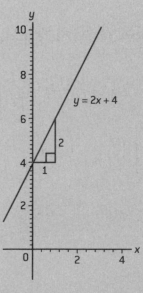

**Figure 12**

**Are they laws?**

Results obtained from experiments are **empirical results**. Carry out the same experiments over wide ranges of pressure or temperature and you would see that the results no longer hold. They are not laws in that sense at all.

There is also a wide meaning to the term "law". Do we mean a scientific law, a law in the legal sense, a law in the sense of a rule for a game (the laws of cricket, for instance), or a mathematical law (Pythagoras's theorem)?

These are all questions for your Theory of knowledge lessons.

## The relationships between temperature, pressure and volume of gas

Experiments on a fixed mass of an ideal gas show that:

| | **Constant pressure** | **Constant volume** | **Constant temperature** |
|---|---|---|---|
| Observation of macroscopic effect | When the pressure of the gas is constant, the graph of volume against temperature is a straight line<br><br>so $V \propto T$ | When the volume of the gas is constant, the graph of pressure against temperature is a straight line<br><br>so $p \propto T$ | When the temperature of the gas is constant, the pressure is inversely proportional to the volume<br><br>so $p \propto \dfrac{1}{V}$<br>or $pV = \text{constant}$ |
| Explanation in microscopic terms | If the pressure stays constant, the ratio of force/area must also be constant. When the volume of the container increases then the area of the walls must also increase. To ensure that pressure and therefore the force/area ratio stay constant, the force on the walls must increase. The only way this can happen is for the speed of the particles to increase so that the momentum exchange at the wall increases too. | When the volume is constant, the wall area does not change so that when the pressure increases, the force on the walls must have increased, and so both the average speed of the particles and, therefore, the temperature must have risen. | When the temperature is constant, the average speed of the particles does not change. When the volume increases, the distance between the walls also goes up. Even though the momentum change is constant, the average time between the particles hitting the wall increases as they have farther to travel between walls. |

**Worked example: Boyle's law**                                                                          **WE**

7.  A sample of a gas at a pressure of 50 kPa has a volume of 25 cm³. The temperature of the gas is held constant.

    Calculate the pressure of the gas when the volume is

    **a)** halved

    **b)** tripled.

*Solution*

**a)** As $p$ and $V$ are inversely proportional at constant temperature, their product ($pV$) is a constant value. If $p$ decreases, $V$ will increase to give the same value of $pV$, and vice versa. Initially in this scenario:

$pV = 50 \text{ kPa} \times 25 \text{ cm}^3 = 1250 \text{ kPa cm}^3$

To find the new value of pressure, simply divide the constant $pV$ by the new value of volume. Since $pV$ is constant, you don't need to convert $p$ and $V$ to SI units as long as you use them consistently for a given problem, unless you are instructed otherwise. In the case of (a), the volume is halved:

$$V_{new} = \frac{25}{2} = 12.5 \text{ kPa}$$

$$p_{new} = \frac{pV}{V_{new}} = \frac{1250}{12.5} = 100 \text{ kPa}$$

b) $V_{new} = 25 \times 3 = 75 \text{ kPa}$

$$p_{new} = \frac{pV}{V_{new}} = \frac{1250}{75} = 17 \text{ kPa}$$

### Question

2 A fixed mass of a gas occupies 0.36 m³ at a pressure of 1.0 atm. Calculate the volume of the gas at a pressure of 2.5 atm when the temperature does not change.

3 A fixed mass of gas has a volume of 1.56 litre. Calculate the volume when the pressure of the gas triples without a change in temperature.

4 A volume of 520 cm³ of a gas is collected at a pressure of 740 mm of mercury. Calculate the volume at standard atmospheric pressure (101.325 kPa).

## 3.3 Changing temperature and state

The first section of this chapter dealt with changes of state and changes in temperature from a qualitative standpoint. Now we turn to a quantitative treatment of the transfer of energy during these changes.

### Changes of state

Imagine 100 g of ice at −10°C in a container. We are going to transfer 1000 J of energy to the ice every second with a 1 kW heater and measure the temperature of the water (in whatever state it exists) as time goes on. The graph of the variation of temperature with time that results from this experiment is shown in figure 13.

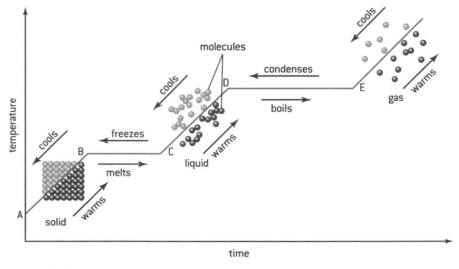

**Figure 13.** Phase changes

The graph shows two regions where the temperature does not change; these correspond to the melting of the ice (BC) and the boiling of the water (DE). During these times, energy is entering the water at a uniform rate but adding to the store of potential energy. The energy goes into breaking the bonds when a liquid boils rather than into increasing the kinetic energy of the molecules. We say that, while the substance is changing state and there is no temperature change, *latent heat* is being transferred into the system. The word "latent" comes from the original English term which itself is derived from the Latin for "remaining hidden".

No change of state occurs in those parts of the graph where the temperature is changing (AB, CD, beyond E). Almost all the energy is being transferred into the kinetic energy of the particles. The rate at which an object increases in temperature as energy is supplied is called its *heat capacity*.

## Specific latent heat

The greater the mass of a substance, the more latent heat is required to change its state. A quantity known as the *specific latent heat* is defined that eliminates the dependence on mass.

It is important to be clear about the state into which a substance is changing; the table shows the terminology used.

| Change | Specific latent heat of ... | |
|---|---|---|
| solid to liquid | ...fusion (melting) | These have the same numerical value for the same substance |
| liquid to solid | ...freezing | |
| liquid to gas | ...vaporization (boiling) | These have the same numerical value for the same substance |
| gas to liquid | ...condensation | |

## Practical skills: Estimating the specific latent heat of melting for water

This is a straightforward experiment that uses a low-voltage laboratory immersion heater.

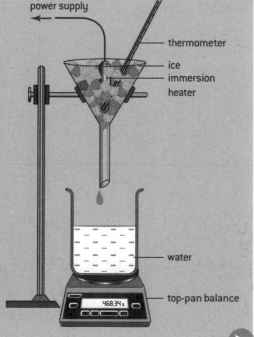

power supply

thermometer

ice

immersion heater

water

top-pan balance

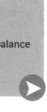

468.34 g

> **Key term**
> The energy required to change the temperature of an object by 1 K is known as the **heat capacity** or the **thermal capacity**.

> **Key term**
> **Specific latent heat** is the energy required to change the state of one kilogram of the substance.
>
> If a substance of mass $m$ changes state with no change in temperature when an energy $Q$ is transferred to it, then the specific latent heat $L$ is given by
> $$L = \frac{Q}{m}$$
> The units of specific latent heat are J kg$^{-1}$. In terms of fundamental units, this is m$^2$ s$^{-2}$.

Pack a filter funnel full of ice at the melting point. Embed the heater in it and connect the power supply. Turn on the supply and note:

- the time, $t$
- the pd, $V$ and current, $I$.

Zero (tare) the balance.

After some time, read the time and the mass of the water.

Repeat the experiment as a control for the same time but with the power supply turned off.

The electrical energy supplied is $VIt$. The mass collected due to the electrical energy (as opposed to energy that is transferred from the room) is $m$. You will need to calculate $m$ from your two mass values.

Use $L = \dfrac{VIt}{m}$ to compute the specific latent heat of melting for the water.

## Worked example: Calculating the specific latent heat of vaporization

8. An electric heater of power 50 W boils a liquid in a pan for 300 s. During this time the mass of the liquid decreases by 39 g. Calculate the specific latent heat of vaporization of the liquid.

*Solution*

The energy supplied by the heater in 300 s = $50 \times 300 = 15\,000$ J

Using $Q = mL$, $15\,000 = 3.9 \times 10^{-2} \times L$; so $L = \dfrac{15\,000}{3.9 \times 10^{-2}}$

$= 3.9 \times 10^{5}$ J kg$^{-1}$

## Specific heat capacity

The amount of energy transferred to a substance to change its temperature depends on both the mass of the substance and the temperature change required. Again, we need a quantity that depends only on the substance itself and not on its mass and temperature change; this is its *specific heat capacity*.

## Practical skills: Estimating the specific heat capacity of a metal cylinder

The method uses an electric immersion heater that slots into a hole drilled into the cylinder. There is another hole for a mercury-in-glass thermometer. (If you do not have access to this, use the method of mixtures that is outlined in Worked example 9.)

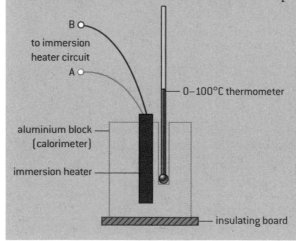

- Measure the mass of the cylinder. As in the specific latent heat experiment, you will need to design an electrical circuit that will allow you to measure the pd across, and the current in, the heater.
- Start a clock, measure the starting temperature of the block and record the pd and current values.
- It is helpful to begin the experiment when the block is a few degrees below room temperature and to end it when the block is the same number of degrees above room temperature. Think about why this is good experimental design.
- At the end, record the time and the final temperature—allow a few minutes after the current is turned off for the temperature to rise to its final value. Do not include this time in the energy calculation.
- The energy supplied to the block (and to the heater and thermometer—these are errors in the measurement) is $V \times I \times t$. This equals the energy involved in the heat capacity change: $m \times c \times \Delta\theta$ where $m$ is the mass of the block and $\Delta\theta$ is the overall change in temperature of the block.

  So $c = \dfrac{VIt}{m \times \Delta\theta}$.

---

**DP ready** | **Nature of science**

**Conservation laws**

The basis of this experiment is that the energy involved in the transfers is **conserved**. We assume that the energy in one transfer is equal to the energy in another. Provided that there is no mass change, this is a correct assumption: another example of a conservation law in physics.

**Internal link**

For an explanation of mass change, see the section on nuclear fusion in **6.2 Energy resources**.

---

**Worked example: Calculating specific heat capacity**

9. Some pieces of copper are heated in a furnace until they reach a high temperature (below the melting point of copper). The copper is then transferred to a beaker of water in an insulated container so that no energy is lost from the copper–water system. The water reaches a steady, higher temperature. (This technique is known as the **method of mixtures**.)

   Calculate, using the data below, the specific heat capacity of the water.

   specific heat capacity of copper = 390 J kg$^{-1}$ K$^{-1}$
   mass of copper = 0.24 kg
   mass of water = 0.51 kg
   final temperature of water = 54°C
   initial temperature of water = 15°C
   temperature of furnace = 950°C

*Solution*

The energy supplied by the copper = the energy gained by the water, so

$m_{Cu}c_{Cu}\Delta\theta_{Cu} = m_{water}c_{water}\Delta\theta_{water}$. Therefore $c_{water} = \dfrac{m_{Cu}c_{Cu}\Delta\theta_{Cu}}{m_{water}\Delta\theta_{water}}$

Substituting for the data gives $c_{water} = 4200$ J kg$^{-1}$ K$^{-1}$

Because the calculation only uses temperature **differences** there is no need to convert the temperatures to kelvin for the calculation.

### The large specific heat capacity of water

Relatively large energies are required to change the temperature of a mass of water. This is of considerable advantage to animals (like us), as our bodies contain large percentages of water. This means it is easy for an animal to keep its temperature constant as large energies only cause small temperature changes.

Often heat capacity and latent heat changes are combined in the same calculation. These are usually straightforward, but you should carry the calculation out in an organized way and present the calculation carefully so that all your steps are clear.

## Worked example: Calculations involving heat capacity and latent heat changes

10. Ice with a mass of 0.15 kg at −5°C is placed in 0.57 kg of water at 34°C. All the ice melts. What is the final temperature $T$ in degrees Celsius of the water and ice?

| | |
|---|---|
| specific heat capacity of ice | = 2100 J kg$^{-1}$ K$^{-1}$ |
| specific heat capacity of water | = 4200 J kg$^{-1}$ K$^{-1}$ |
| specific latent heat of fusion of ice | = $3.34 \times 10^5$ J kg$^{-1}$ |

*Solution*

There are four parts to the calculation:

| | |
|---|---|
| • energy gained by ice warming to 0°C | = $0.15 \times 2100 \times 5$ |
| • energy gained by ice in melting | = $0.15 \times 3.34 \times 10^5$ |
| • energy gained by ice (as water) in warming to final $T$ | = $0.15 \times 4200 \times (T - 0)$ |
| • energy lost by (original) water in cooling to final $T$ | = $0.57 \times 4200 \times (34 - T)$ |

Based on the law of conservation of energy, the energy gained by the ice (the first three calculations) is equal to the energy lost by the water (the last calculation). The first three calculations sum to $51675 + 630T$ and the last calculation is equal to $81396 - 2394T$. We can then equate these expressions to find $T$:

$51675 + 630T = 81396 - 2394T$

$3024T = 29721$

$T = 9.8°C$

## Maths skills: Solving linear simultaneous equations

Worked example 10 involves solution of two linear simultaneous equations (equations that contain the same unknowns). In the example, the unknowns are $T$ and $34 - T$, and the second unknown is already expressed simply in terms of the first. More generally, such equations can be solved as shown below.

Suppose the two equations are $3x + 2y = 7$ and $2x + 3y = 8$.

Multiply both equations to give the $x$-coefficient the same number. In this case, multiply the first equation by 2 and the second by 3:

$6x + 4y = 14$ and $6x + 9y = 24$

Then subtract one equation from the other so that the first term disappears

$(6x - 6x) + (4y - 9y) = 14 - 24$

or $-5y = -10$   *Keep track of the signs!*

Therefore $y = \dfrac{-10}{-5} = +2$

And, substituting $y$ into the first equation gives $3x + 4 = 7$,

so $3x = 7 - 4 = 3$, $x = 1$

**Question**

5  An immersion heater with a power output of 2.0 kW is placed in an insulated beaker of water. The water boils 120 s after the heater is switched on.

Calculate, using the data below, the specific heat capacity of the water. Ignore any energy absorbed by the beaker.

| | |
|---|---|
| mass of empty beaker | = 0.035 kg |
| mass of beaker with water | = 0.750 kg |
| initial temperature of water | = 20°C |

One last point about these energy changes. Think about the example of the ice melting into the water, the ice warming up, the water cooling down. We assumed that the energy was kept in the system. But what happens in practice? We all know that leaving a hot drink standing for a while means that it will no longer be a hot drink, and we either drink it cold or reheat it. This energy has been **dissipated** into the environment. It is difficult, if not impossible, to recover. This is the fate of all energy.

As you progress through the Physics Diploma Programme (especially if you undertake Option B) you will realize that sources of energy are gradually being used to do work, and that in the process the energy becomes "smeared out" to a uniform temperature. It has been suggested that one plausible end for the whole universe is a heat death. The universe will eventually approach an equilibrium where everything is at the same low temperature and no further energy transfers are possible.

**Chapter summary**

Make sure that you have a working knowledge of the following concepts and definitions:

☐ The three normal states of matter are solid, liquid and gas. These are characterized by the differences in particle motion between the states. All these states are made up of atoms, ions and molecules.

☐ Scientific models can take a microscopic or a macroscopic view of matter. Both are equally valid but bring out distinct aspects of materials.

☐ Solids have particles that are fixed. The particles in liquids are largely fixed in position but can slide past each other. The particles of a gas are completely independent.

☐ Temperature is a measure of the average kinetic energy of the particles.

☐ Evaporation and boiling are both examples of a liquid changing phase to the gaseous state.

☐ Temperature scales require at least two fixed points. The kelvin scale and the Celsius scale are commonly used in science. The kelvin scale is a theoretical scale closely related to the properties of an ideal gas.

☐ Pressure can be defined for each state of matter, but is central to the description of the state of a gas.

☐ The pressure in a liquid is directly proportional to the length of a column of the liquid and depends on its density.

☐ Solids can exert a pressure on a surface given by $\dfrac{\text{weight force of the solid}}{\text{area of solid in contact with surface}}$.

☐ Pressure measuring devices include the U-tube manometer, the mercury barometer and the Bourdon gauge.

☐ The pressure of a gas arises from the transfer of momentum from the moving particles of the gas to the walls of its container.

- There are three rules that connect the pressure $p$ of a fixed mass of gas, its volume $V$ and its temperature $T$ (in kelvin). These are:
  - $pV$ = constant when $T$ is constant
  - $p \propto T$ when $V$ is constant
  - $V \propto T$ when $p$ is constant
- Latent heat is the name given to the energy required to change the state of a substance. Specific latent heat is the amount of energy required per kilogram of substance to change its state.
- Specific heat capacity is the amount of energy required per kilogram to change the temperature of a substance by one kelvin when there is no change of state.

## Additional questions

1. Two identical solid blocks X and Y are placed in contact. X is at a higher temperature than Y.
   a) Compare the motion of the particles in X with the motion of those in Y.
   b) Discuss the transfer of thermal energy between X and Y.
2. An ideal gas is held in a container with fixed walls. Explain, with reference to the kinetic theory, why
   a) the gas exerts pressure on the walls
   b) the pressure increases when the temperature increases.
3. a) Explain the process of condensation.
   b) Explain why a glass mirror in a bathroom "steams up".
4. Explain why
   a) the foundations of a house wall are wider than the wall itself
   b) agricultural vehicles usually have very wide tyres
   c) boots and shoes for sport often have spikes or studs.
5. A brick has the dimensions 35 cm × 25 cm × 18 cm. The mass of the brick is 15 kg. Calculate
   a) the weight of the brick
   b) the greatest pressure the brick can exert on horizontal ground
   c) the least pressure the brick can exert on horizontal ground.
6. Calculate the length of a barometer tube when water is used in place of mercury to measure atmospheric pressure. The density of mercury is 13 600 kg m$^{-3}$ and the density of water is 1000 kg m$^{-3}$.
7. Explain why weather-forecasting balloons that go into the high atmosphere are not completely filled with helium gas before release.

8. A beaker containing trapped air at atmospheric pressure is placed upside down on the surface of a swimming pool.

   Estimate the volume of the air in the beaker when it is taken to a point 5 m below the surface of the pool. Neglect any changes in temperature.

You may require the following data for the next few questions:

specific heat capacity of water = 4200 J kg$^{-1}$ K$^{-1}$
specific heat capacity of ice = 2100 J kg$^{-1}$ K$^{-1}$
specific latent heat of fusion of ice = 3.3 × 10$^2$ kJ kg$^{-1}$
specific latent heat of vaporization of water = 2.3 MJ kg$^{-1}$

9. In an ice-making machine, water enters the machine at 15°C and is frozen to form ice cubes at −5°C.

   a) Determine the energy that must be transferred from 1.0 kg of water to form 1.0 kg of ice cubes.

   b) The machine can freeze up to 2.5 kg of ice every 300 s. Calculate the rate at which energy must be removed by the machine.

10. A cup containing 150 g of water at 18°C is boiled by bubbling steam at 100°C through the water.

    Determine the minimum mass of water that will be in the cup when the liquid reaches the boiling point.

11. A thermometer of mass 100 g has a specific heat capacity of 13 J kg$^{-1}$ K$^{-1}$. The thermometer, initially at 20°C, is used to measure the temperature of 50 g of water. The thermometer measures the temperature of the water to be 36°C.

    a) Calculate the temperature change of the water as a result of introducing the thermometer.

    b) Explain why the thermal capacity of a thermometer should be small and the factors that need to be considered to achieve this.

12. A liquid in a flask is boiling using an immersion heater. When the heater is transferring 100 W to the liquid, 0.100 g of liquid boils each second. When the heater power is increased to 180 W, 0.200 g of liquid boils every second.

    Calculate:

    a) the specific latent heat of the liquid

    b) the rate at which thermal energy is transferred by the flask.

# 4 Waves

> " The discovery of electrical waves has not merely scientific interest ... it has had a profound influence on civilization; it has been instrumental in providing the methods which may bring all inhabitants of the world within hearing distance of each other and has potentialities social, educational and political which we are only beginning to realize. "
>
> **Sir Joseph J Thomson, on James Clerk Maxwell's discovery of electromagnetic waves, *James Clerk Maxwell: A Commemorative Volume 1831–1931* (1931)**

## Chapter context

**Waves** are important throughout physics, from the simple example of a water wave through to a model of the behaviour of the **electron**.

In this chapter you will look at the properties of all types of waves and then in detail at the physics of **light** and of **sound**. We also consider **electromagnetic** waves in general these are of great importance to us when we send information from one person to another.

## Learning objectives

In this chapter you will learn about:

→ displaying **wave motion**: waves and rays

→ **transfer of energy** in wave motion

→ the **wave-speed equation**

→ **displacement–distance** and **displacement–time graphs** for wave motion

→ **diffraction**

→ **radiation intensity** and the **inverse square law**

→ **refraction** and the **reversibility** of light

→ **total internal reflection** and **optic fibres**

→ **electromagnetic** waves

→ **echoes** and other phenomena of **sound waves**.

## 🔑 Key terms introduced

→ transverse and longitudinal waves

→ displacement and amplitude of waves, troughs and crests

→ time period and frequency

→ wavelength

→ compression and rarefaction

→ reflection

→ refraction

→ diffraction

→ real and virtual images

→ refractive index

---

 **DP link**

You will learn about the basic properties of waves when you study **4.2 Travelling waves** in the IB Diploma Programme Physics course.

## 4.1 Waves in theory

### Wave motion

Skim a pebble along the still surface of a lake or pond and ripples appear at the points where the pebble touches the water (figure 1). The ripples move outwards as circles, eventually fading. These are mechanical **surface waves** on the water.

Waves can be divided into two main types:

- **Mechanical waves** need a **medium** – a material – to carry them. They cannot travel through a vacuum. Examples include waves in water, waves on stretched strings and sound waves in a gas.

- **Electromagnetic waves** can, in principle, travel through solids, liquids or gases, and, important to us, the vacuum of space.

A common feature of all waves is that they transfer energy from one place to another but leave the medium unchanged once the wave has passed. When the waves on the water surface have arrived at the lake's edge, the surface will be still again. All the original kinetic energy of the stone has disappeared into heating the air, the water and the stones at the edge.

## Transverse waves

Figure 2 shows a Slinky spring – a familiar children's toy that can illustrate many wave properties. Stretch the Slinky out on a long table or the floor and move one end sideways at right angles to the length. An obvious pulse will move along the spring. When this is repeated continuously then a wave moves along the Slinky, disappearing at the other end. This is a *transverse* wave.

**Figure 1.** Ripples on a pond

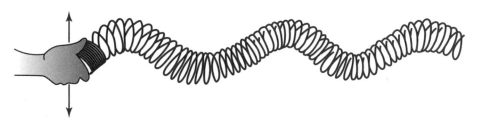

**Figure 2.** Waves on a Slinky spring

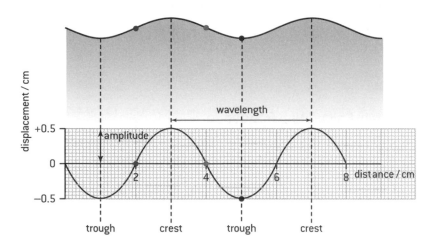

**Figure 3.** Transverse wave profile

A photograph of the Slinky will look something like the shape in figure 3. At the top of figure 3 is the spring itself. Below it is a graph of the variation of *displacement* from the rest position with distance along the Slinky (known as a **displacement–distance** graph).

The maximum displacement of the Slinky is the *amplitude* of the wave and is the same as the maximum distance travelled by the hand moving the Slinky from the central position.

## Displacement–distance graphs

There is a need to be clear about directions here. The graph $y$-axis is marked " + " and "–". In the case of figure 3, " + " means up and "–" means down, but for another wave, displacement might be to left and right. Notice that the amplitude is measured from the middle of the oscillation—the equilibrium position—to the maximum (or minimum), **not** from minimum to maximum (top to bottom).

### Key term

The **displacement** is the distance through which a point on the wave has moved from its rest (zero or equilibrium) position.

The **amplitude** is the maximum distance moved by a point from the rest position.

We call the minimum position on the displacement–distance graph the **trough** and the maximum the **crest**.

Other important features of a wave include:

- its recurrence at regular intervals of time and regular distances: these regular changes arise from the repetitive way in which, for example, the Slinky is moved to generate the wave
- its *cycle*—the motion of a point on the wave from the equilibrium position to the maximum amplitude, back through the centre and in the opposite direction, then returning to the equilibrium position
- the constant wave *frequency*—the number of cycles generated every second
- its constant *time period*.

 **Key term**

A **cycle** is the movement of a point on the wave through one complete oscillation, from the rest position to the maximum amplitude, to the opposite maximum amplitude, then back to the equilibrium position.

**Frequency** is the number of cycles completed in one second.

The unit of frequency is the hertz (abbreviated Hz), named for Heinrich Rudolph Hertz, a German physicist from the later 19th century. The fundamental (base) SI unit is s$^{-1}$.

The **time period** is the time taken for one complete cycle.

The relationship between time period $T$ and frequency $f$ is given by:
$$T = \frac{1}{f}.$$

**Worked example: Calculating frequency**

1. A student uses a Slinky to generate a wave. The student takes 3.2 s to move the Slinky through one cycle. Calculate the frequency of the wave.

*Solution*

$T = 3.2$ s and so $f = \dfrac{1}{T} = \dfrac{1}{3.2} = 0.31$ Hz

 **Key term**

**Wavelength** $\lambda$ is the shortest distance between two points that are in phase on the wave, for example two consecutive crests or two consecutive troughs.

Wavelength is a distance, so the unit of wavelength is the metre.

**Wave speed** is the distance moved forward every second by the wave, for example how far the crest moves in one second.

### Speed, frequency and wavelength

Look at the graph in figure 3 again. Notice that the distance between the two maxima (crest to crest) on the curve is 4.0 cm. This distance is known as the *wavelength* $\lambda$ of the wave.

Frequency and wavelength are two crucial pieces of information about any wave or oscillation and can be combined to yield the **speed** $c$ of the wave. Wave speed is no different from the speeds described in *1.1 Faster and faster*; however, it can be expressed in terms of frequency and wavelength.

When one part of the wave, say the crest, has moved a distance $\lambda$, then time $T \left( = \dfrac{1}{f} \right)$ has passed.

So the speed $c = \dfrac{\text{distance moved forward}}{\text{time taken for movement}}$

$$= \frac{\lambda}{T} = \frac{\lambda}{\left(\dfrac{1}{f}\right)} = f\lambda.$$

---

**DP ready** **Nature of science**

### The wave-speed equation

This equation, $c = f\lambda$, is called the wave-speed equation.

Frequency has the units $s^{-1}$; wavelength has the unit m. So wave speed has the unit $m \times s^{-1}$, which is $m\,s^{-1}$—the same as that for speed discussed in *1 Motion and force*.

---

**DP ready** **Theory of knowledge**

### Huygens versus Newton

The road to a consistent scientific theory is not always smooth. In the 17th century Newton supported the view of French philosopher Descartes that light is made of small, separate particles, then called "corpuscles", that travel in a straight line at a definite speed. Christiaan Huygens, a Dutch scientist, took the view that light consists of waves. Huygens's view eventually prevailed on the available evidence—for a time. But in the 20th century physicists adopted the view that electromagnetic radiation has a dual nature: wave and particle. The energy packets that make up what we call light are now known as "photons". Photons possess energy and momentum, but no mass. This tells us about the importance of the concept of momentum and its conservation.

---

### Worked example: Calculating wavelength

2. A sound of frequency 440 Hz is produced by a violin. The speed of sound in air is $340\ m\,s^{-1}$. Calculate the wavelength of the violin note in the air.

*Solution*

$c = f\lambda$ so $\lambda = \dfrac{c}{f} = \dfrac{340}{440} = 0.77$ m

---

### Question

1 Complete the following table.

| | Sound wave in air | Water wave | Electromagnetic wave | Wave on a string | Ultrasound in water |
|---|---|---|---|---|---|
| Frequency (Hz) | | 0.070 | | 260 | $2.9 \times 10^4$ |
| Wavelength (m) | 0.20 | 340 | $6.5 \times 10^{-7}$ | 1.6 | $5.2 \times 10^{-2}$ |
| Speed ($m\,s^{-1}$) | 340 | | $3.0 \times 10^8$ | | |

## Key term

In a **transverse** wave, the displacement is at 90° to the direction in which the energy moves in the medium.

In a **longitudinal** wave, the energy transfer and the motion of the particles are in the same direction.

## Longitudinal waves

So far, we have defined the quantities of wave motion using the example of a *transverse* wave. This is a wave in which the displacement is at 90° to the direction in which the energy moves. In the case of the Slinky, the energy moves along the spring and the displacement of the coils is in a perpendicular direction.

There is a different relationship between the direction of energy transfer and the displacement for *longitudinal* waves. Figure 4 shows the motion of a Slinky when a longitudinal wave passes along it. Here the energy transfer and the motion of the coils are in the **same** direction. Every individual coil in the spring moves backwards and forward (left to right and back again as you look at the picture).

rarefaction     compression

**Figure 4.** Longitudinal waves on a Slinky spring

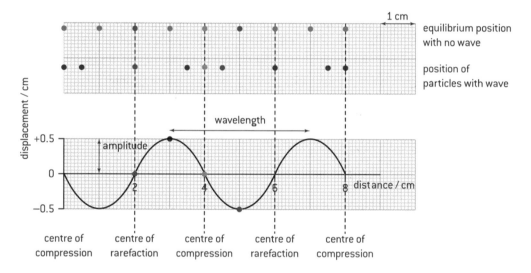

**Figure 5.** Longitudinal wave profile

This wave type too can be analysed with a displacement–distance graph (figure 5). This time, for a Slinky, the displacement is **along** the coil, not at right angles to it. The graph needs more careful interpretation than the equivalent graph for the transverse wave. The *y*-axis shows displacement along the axis of the Slinky; a positive value on the *y*-axis means movement to the right and a negative value means movement to the left.

Now focus on the upper part of figure 5. This shows nine particles, equally spaced before the wave goes through the Slinky. The graph tells us the displacement of each of these particles at one instant. The odd numbered particles (counting from either end) are not displaced from their equilibrium position because the graph of displacement–distance indicates that the displacement is zero. Even-numbered particles are displaced either to the right or to the left by 0.5 cm (the value of the graph at their position). The changes in position are shown in the lower of the two rows of particles. In some places the movement has pushed the particles together (a *compression*); in other places they are moved apart (a *rarefaction*). Compare this diagram with figure 4: the Slinky coils have a pattern of compression (coils moved together) and rarefaction (coils moved apart).

 **Key term**

A longitudinal wave consists of regular **compressions** (particles close together) and **rarefactions** (particles spread apart).

## Displacement–time graphs

The displacement–distance graph gives a "snapshot" of the whole wave at one instant, but gives little information about the motion of a particular point on the wave with time. For that a different graph is needed, the **displacement–time** graph (figure 6). This graph looks very similar to the displacement–distance variant, so care is needed.

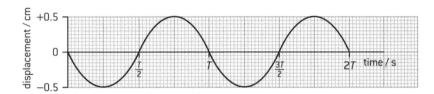

**Figure 6.** Displacement–time graph for a wave

We shall consider a transverse wave on a spring to explain displacement–time graphs but, again, the graph can apply to any type of wave. Figure 6 shows the behaviour of **one** coil of the spring as a wave two wavelengths long moves through the coil position. Assume that the overall movement of the wave is from the left side of the page to the right—the displacement–time graph cannot tell us this as we are focusing on only one coil. The graph shows that the displacement of the coil from its rest position is 0 at time 0, just before the wave on the Slinky has reached this particular coil. When the wave reaches it, the coil moves downward, then from its maximum displacement (amplitude) it returns to the rest position at time $\frac{T}{2}$, carries on upward to its greatest displacement in that direction and reverses back down toward the rest position, which it passes at time $T$, one time period. The cycle is repeated a second time on the graph.

The displacement–time graph gives information about the amplitude (as before) and the time period $T$, but cannot give direct information about the wavelength. However, when you can see two displacement-time plots for different coils on the same graph you can deduce the wavelength if you know their distance apart on the Slinky (see Worked example 3).

### Worked example: Displacement–time graphs

3. Figure 7 is the displacement–time graph for two coils on the same spring as a wave moves along it. The coils are 0.40 m apart. Calculate:

   **a)** the frequency of the wave

   **b)** its wavelength.

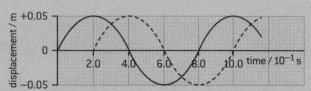

**Figure 7.** Displacement–time graph for two points in the same wave

*Solution*

**a)** Each coil goes through one complete oscillation (cycle) in $T = 0.80$ s, so $f$ is $\dfrac{1}{0.8} = 1.25$ Hz.

**b)** The second coil undergoes the same displacement 0.20 s later than the first coil and is 0.40 m from it, so the speed on the spring is $\dfrac{0.40}{0.20} = 2.0$ m s$^{-1}$.

Using $\lambda = \dfrac{c}{f} = \dfrac{2.0}{1.25} = 1.6$ m.

### Waves and rays

Springs and stretched strings are easy to visualize because their wave shapes and the graphs of their motion look similar. Other waves, such as invisible sound waves in air or light waves, are less easy. Wavefronts and rays are one way to help us represent waves of all types (figure 8).

- A **wavefront** is a surface that travels with the wave. You can think of this as a single wave crest moving along. When there is more than one wavefront (for example, consecutive crests following each other a wavelength apart) then we get a good sense of how the wave is moving and how the wave shape is changing.

- A **ray** shows the direction in which a wave transfers energy. This direction is perpendicular to the wavefront.

- Wavefront diagrams combined with rays give information about the changes in the wave over time. Figure 8(a) shows a set of parallel wavefronts together with the ray. The wavefronts move to the right and are equally spaced, implying that the wavelength is not changing. Figure 8(b) shows wavefronts that have begun from a point source. We can tell the age of each circular wavefront from its radius of curvature.

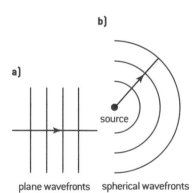

plane wavefronts    spherical wavefronts

**Figure 8.** Wavefronts and their associated rays: **(a)** parallel waves; **(b)** point source

## Practical skills: Using a ripple tank

A **ripple tank** (figure 9) is often used to make wave phenomena visible. Light from above the tank is incident on the water layer. Crests on the water surface focus the light, troughs defocus it, casting shadows on the screen below the tank.

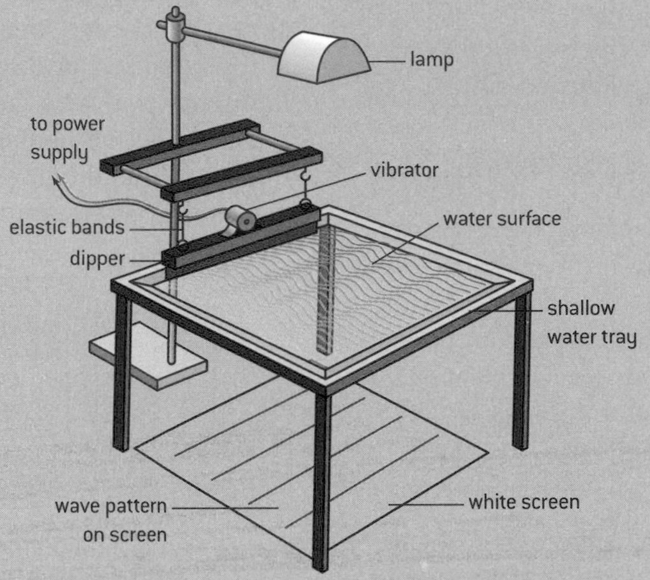

lamp

to power supply

vibrator

elastic bands

water surface

dipper

water surface

shallow water tray

wave pattern on screen

white screen

**Figure 9.** A ripple tank

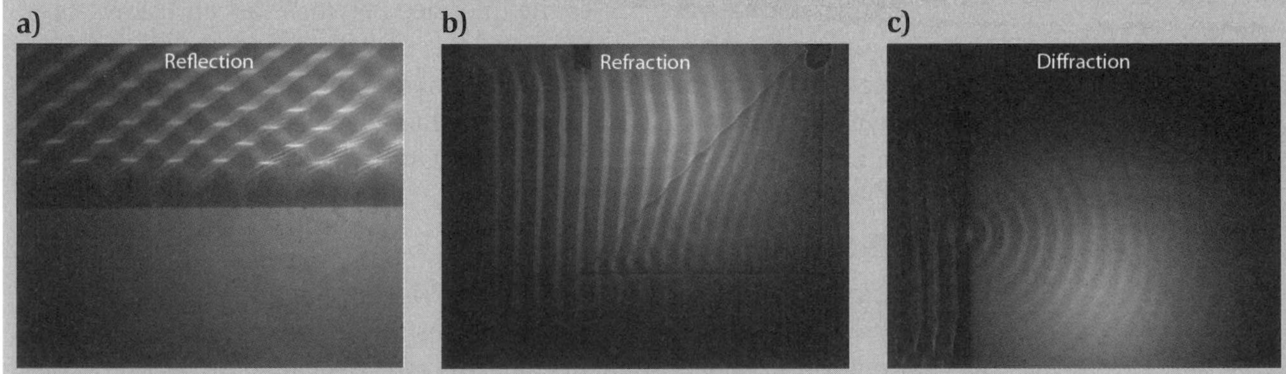

a)

Reflection

b)

Refraction

c)

Diffraction

**Figure 10.** Interpreting ripple-tank images

- Figure 10(a) shows reflection in a ripple tank. The waves are coming in from the top left of the image (diagonal lines) and are reflected towards top right.
- Figure 10(b) shows refraction. A transparent sheet in the right-hand side of the tank with a diagonal division bottom left to top right makes the waves change direction.
- In figure 10(c) plane waves from the left are diffracted at a small gap in a barrier, forming semi-circular waves to the right of the barrier.

## Reflection

Figure 11 shows the ripple-tank pattern for *reflection*. Waves are incident from the left onto a flat metal surface placed in the tank. The waves reflect from the metal and move off at a different angle to the original direction. A **normal** (a line drawn at 90°, shown here as a dashed line) is constructed to the metal surface. The rays that correspond to the two sets of wavefronts are also included, though of course these are not seen in the photographs. When accurate measurements are made it is found that the angle between the incident ray and the normal (called the *angle of incidence*—$i$ in the diagram) is the same as the angle between the reflected ray and the normal (the *angle of reflection*—$r$). This rule always applies when waves are reflected from a plane (flat) surface; it is one of the *laws of reflection*.

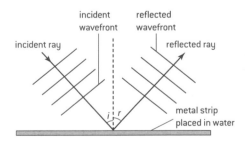

incident wavefront    reflected wavefront

incident ray    reflected ray

$i$   $r$

metal strip placed in water

**Figure 11.** Wavefront and ray diagram for reflection

## Internal link

The laws of reflection and the laws of refraction are stated in **4.2 Physics of light**.

## Key term

**Reflection** is the redirection of waves from a surface such that the angle between the incident ray and the normal (the **angle of incidence**) is equal to the angle between the reflected ray and the normal (**angle of reflection**).

In **refraction** rays change direction because the wave speed changes when the wave passes from one medium to another.

In **diffraction** plane wavefronts bend as the wave energy passes through a gap or around an obstacle.

**Figure 13.** Diffraction around an edge

## DP link

You will learn about the effects of diffraction and interference in
**4.4 Wave behaviour** and in the AHL topics
**9.2 Single-slit diffraction**,
**9.3 Interference** and
**9.4 Resolution** on the DP Physics course.

## Refraction

When waves pass from deep water to shallow water (figure 12) they slow down (wave speed depends on water depth). This also happens at the boundary or **interface** between media. Since $c = f\lambda$ and the frequency is always the same, the wavelength in shallow water must be shorter. Look carefully at the wavefront separation in both figure 10(b) and figure 12 and remember that the wavefronts are a wavelength apart. You will see that the wavelength changes.

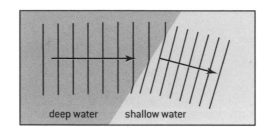

deep water     shallow water

**Figure 12.** Wavefront and ray diagram for refraction

The direction of the wavefronts (and therefore the ray direction) also changes as the wave passes the interface between fast and slow. Rays changing direction as speed changes is called *refraction*. It occurs because the wavefront is continuous across the interface. Successive wavefronts must have different spacings (wavelengths) each side of the boundary, and the only way this can happen is for the direction of the wavefronts to swing round.

## Diffraction

A curious effect happens when plane wavefronts are incident on a narrow gap. As the wave energy passes through the gap, the plane wavefronts become semicircular. This effect is called *diffraction*. The wavelength does not change in this case (because there is no change in wave speed).

As you might expect, when the gap is replaced by an edge the pattern changes to leave just half of the diffraction (figure 13).

## Amplitude and intensity

In *1.3 Work and energy* and *2.1 Electric fields and currents* the kinetic energy associated with moving systems and energy transfer in electric circuits were discussed. It is also relevant to ask about the energy associated with a wave. To do this we can draw parallels with the mechanical and electrical cases.

You have learnt that the kinetic energy $E_k$ of a mass $m$ moving with a speed $v$ is $\frac{1}{2}mv^2$ and that the energy transferred in a resistance $R$ with an electrical current $I$ in it is $I^2R$. Notice that in both cases the energy depends on a quantity squared together with another quantity that has the quality of inertia (the mass) or a dissipative effect ($R$).

It seems reasonable, therefore, to suggest that the energy $E$ associated with a wave depends on the amplitude $A$ of the wave, also squared. In symbols: $E \propto A^2$. Doubling the amplitude of a wave makes the energy four times greater; tripling the amplitude means a nine-fold increase in energy.

Sometimes, a simple argument will help you to remember a piece of theory.

For a spring, when the amplitude of movement is doubled to $2A$, keeping the frequency and speed, and therefore the wavelength, constant, two things must happen:

- Each coil moves a distance of $4A$ from maximum to minimum (crest to trough) compared with $2A$ before.

- Each coil moves twice as fast to cover this increased distance in the same time period (so that the frequency does not change).

This raises the question of how we observe a sound or a light wave. The answer is that the ear and the eye detect **energy** transfers rather than *amplitude* changes. Our eyes and ears are sensitive to the power of a wave (energy arriving per second) as it is spread out over the reception membrane: for hearing this is the ear drum and for vision it is the sensitive retina at the back of the eye. The relevant quantity here is the *intensity* of the wave.

How does the intensity of a wave vary with distance $r$ from the source as a wave spreads out? Suppose a point source of waves transfers energy at a rate $P$ (this is the power of the source), and the energy spreads out evenly. At a distance $r$ from the point, the energy will be distributed over the surface of a sphere with radius $r$. The intensity at the surface of this sphere must therefore be $\dfrac{P}{4\pi r^2}$ because $4\pi r^2$ is the surface of a sphere. Figure 14(a) shows the sphere with a small part of the surface defined by four radial lines. Suppose we double the distance from the centre. Figure 14(b) shows what happens as the lines are extended to a distance of $2r$: the original surface area becomes four times as large. Triple the distance and the area is nine times the original area. This confirms that $I \propto \dfrac{1}{r^2}$.

This is an **inverse square law** relationship.

**Key term**

**Intensity** is

$$\dfrac{\text{power } P \text{ arriving at the surface}}{\text{area } A \text{ of the surface}}$$

$$= \dfrac{P}{A}$$

The unit of intensity is W m$^{-2}$.

**a)**

**b)**

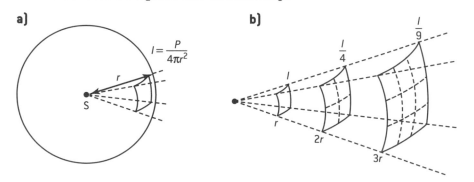

**Figure 14.** The intensity of energy spreading out from a point surface follows an inverse square law

**DP ready**   Theory of knowledge

**Inverse square laws**

The inverse square law relationship is a particularly common one in science when distance is related to another quantity, stemming from the mathematical properties of three-dimensional space. You will meet it in gravitational field theory and when you study electric fields. To what extent is this a true property of space or a by-product of the way we describe it?

## Worked example: Intensity and amplitude

4.  A sound has an intensity of $4.0 \times 10^{-5}$ W m$^{-2}$ at a distance of 25 m from the point source.

    a)  Show that the intensity at 100 m from the point source is $2.5 \times 10^{-6}$ W m$^{-2}$.

    b)  Determine the ratio $\dfrac{\text{amplitude at 25 m}}{\text{amplitude at 100 m}}$ for this sound.

*Solution*

a)  Rearranging the definition of $I$ gives $Ir^2 = \dfrac{P}{4\pi}$. $\dfrac{P}{4\pi}$ is constant for any single source and so for this case $I_{100} \times 100^2 = I_{25} \times 25^2$, so $I_{100} = I_{25} \times \left(\dfrac{25}{100}\right)^2 = 4.0 \times 10^{-5} \times \left(\dfrac{1}{4}\right)^2$, which is the given answer.

b)  $\dfrac{I_{25}}{I_{100}} = \dfrac{A_{25}^2}{A_{100}^2}$ so $\dfrac{A_{25}}{A_{100}} = \sqrt{\dfrac{I_{25}}{I_{100}}} = 4$

### Internal link

A question with the command term "show that" needs some care. There is advice on all the command terms in 7.4 **Understanding questions**.

## 4.2 Physics of light

### Reflection

This section examines the behaviour of light when treated as a wave. Although we cannot see the wavefronts of light, we can easily produce light rays experimentally and observe their behaviour.

Observations made with a ripple tank suggest that when a ray is incident on a reflecting plane (flat) surface, then the angle of incidence and the angle of reflection are equal.

### Key term

**Laws of reflection**

1.  The incident ray, the reflected ray and the normal to the surface of the mirror all lie in the same plane.
2.  Angle of incidence $i =$ angle of reflection $r$ (figure 15).

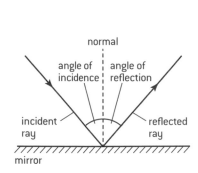

**Figure 15.** Reflection of one ray

### Practical skills: Working with light—Reflection

You will need a ray box with a single slit to produce one ray, a plane mirror, a drawing board and a piece of paper.

Place the paper on the board and put the mirror on a line drawn in the centre of the paper.

Shine a ray on to the mirror surface. You should see a reflected ray coming from the mirror surface (figure 15). Mark the incident ray and the reflected ray with two crosses each to define their position. You can now remove the mirror.

Use a protractor to construct the normal at the point where the rays meet the mirror and measure the angle of incidence and the angle of reflection. Repeat for a number of different incidence angles. You should find that the laws of reflection are confirmed.

Look into a plane mirror and you see an **image** of yourself. You are acting as the **object** for the mirror. What are the properties of this image? To answer this we need to break the question down into a series of questions:

- Where is the image?
- How big is it?
- What other properties does it have?

The best way to answer these is to make an accurate construction using the laws of reflection for rays as they leave an object and reflect from the mirror.

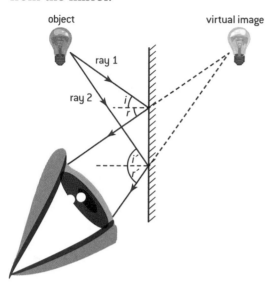

**Figure 16.** Forming an image in a plane mirror

Figure 16 shows two rays from a lamp. They are incident on a mirror at different angles $i$, so have different values for $r$ when they are reflected at the mirror surface, obeying the laws of reflection. After reflection the rays enter an observer's eye at different angles. The observer's brain constructs these rays backwards to determine where they began. They appear to have come from a place behind the mirror. You can confirm from measurements on this page that the **image is the same distance behind the mirror as the object is in front** and the **image is the same size as the object**.

The remarkable thing is that this image formed by the plane mirror does not really exist. It is said to be *virtual* as opposed to *real* (figure 17).

> **Key term**
>
> A **real** image can be formed on a screen, as in a cinema or on the retina of our eyes; a **virtual** image cannot be formed on a screen and needs another lens (which can be the one in our eye) to form a real image.

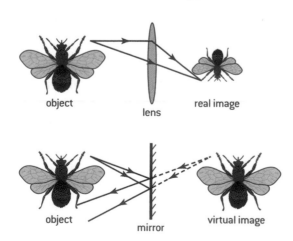

**Figure 17.** Real and virtual images

### Rotating a mirror

An important skill you need to develop is that of designing and carrying out an experiment to answer a question. The question in this case is: when a light ray, fixed in direction, is incident on a plane mirror and the mirror rotates, what happens to the angle of reflection?

This question can be answered either with a thought experiment or a piece of practical work.

If your choice is a practical, you will need to:

- think about the results you require (observations, measurements)
- consider what you want to vary, the measurements you need and the possible variables you need to keep constant
- design the apparatus you require and plan how to analyse the results
- carry out the experiment
- analyse the results
- evaluate the quality of your experiment and decide whether there are refinements worth making before finishing.

### Worked example: Reflection of sound

**WE**

5. X stands 200 m from a wall and whistles a note for a duration of 0.50 s. Y stands 600 m from the wall and hears the sound twice. The speed of sound in air is 340 m s$^{-1}$.

   Determine the time gap between the two sounds.

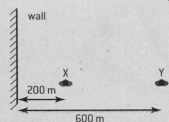

**Solution**

The first (direct) whistle travels a distance of 400 m from X to Y. This takes $\frac{400}{340} = 1.176$ s and the sound ends 0.50 s later at 1.676 s. The second whistle is an echo and travels 800 m to Y. It arrives $\frac{800}{340} = 2.353$ s after the whistle began. The difference between the end of the first sound and the beginning of the next is $2.353 - 1.676 = 0.68$ s.

### Question

**Q**

2  A periscope is a device for seeing over an obstacle when line-of-sight observation is not possible. Use the internet to find the arrangement of mirrors in a periscope.

   Describe the final image as viewed through a periscope.

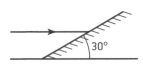

3  A ray of light parallel to the ground is incident on a mirror that has an angle of 30° to the ground.

   Determine:

   a) the angle to the vertical made by the reflected ray, and

   b) the angle through which the mirror must be rotated so that the reflected ray is vertical.

## Refraction

### Practical skills: Working with light—Refraction

The technique used is similar to that in the *Working with light: Reflection* experiment. Use a single ray from a ray box and shine it through the flat side of a semicircular block of glass or plastic. Make sure that the ray enters the block at the centre (figure 18). Draw around the block to mark its position on the paper, and mark the path of the ray through the block. Take measurements to determine the angle of incidence *i* and the angle of refraction *r*. You may also see a weak reflected ray starting where the incident ray enters the block. Vary the angle and look for a pattern in the results.

You should find that *r* is always smaller than *i* and that *r* increases as *i* increases.

Try to verify the mathematical **relationship** stated in the text. You can either carry out the division for each pair of *i* and *r* that you have, or alternatively plot a graph of sin *i* against sin *r*. What shape do you expect this graph to have for the relationship to be confirmed?

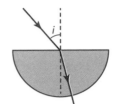

**Figure 18.** Refraction of a single ray

The ripple tank showed *refraction* when the waves travelled from one medium to another with a change in wave speed and the rays and wavefronts changed direction. Experimental results show that the ratio $\dfrac{\sin i}{\sin r}$ is a constant that depends on the material. Change the block for one made from a different substance and the ratio changes.

### Key term

**Snell's law and the laws of refraction**

1. The incident ray, the refracted ray and the normal to the boundary of the media all lie in the same plane.

2. For waves of a given frequency and for a specified pair of media, $\dfrac{\sin i}{\sin r} =$ constant where *i* is the angle of incidence and *r* is the angle of refraction (figure 19). The constant is known as the relative **refractive index**. This is Snell's law.

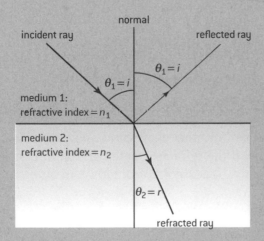

**Figure 19.** Snell's law

**But is it really Snell's law?**

In 1621 a Dutch scientist, Willebrord Snellius (known as Snell in the English-speaking world), stated this law, but the rule had been described at least 600 years before this by Persian scientists. Today the rule is known as Snell's law (or Descartes' law in French-speaking parts of the world).

What other examples are there of scientific rules that have crossed borders and times in this way?

The *refractive index* for light passing from medium 1 to medium 2 ($_1n_2$) is $\dfrac{\sin i}{\sin r}$, also written as $\dfrac{\sin \theta_1}{\sin \theta_2}$. Strictly this is the relative refractive index as it is not referenced to light moving to a medium from a vacuum.

**Absolute refractive index**

Refractive index, as defined before, depends on both media and on other conditions, so the value for glass to air at 20°C is different from that for glass to air at 40°C. The *absolute refractive index n* is defined to remove this problem.

$$n = \frac{\text{speed of electromagnetic waves in a vacuum}}{\text{speed of electromagnetic waves in the medium}} = \frac{c}{v}$$

We can now write Snell's law as $n_1 \sin \theta_1 = n_2 \sin \theta_2$, where $n_1$ and $n_2$ are the two absolute refractive indices.

The absolute refractive index usually depends on frequency and is always greater than 1 (except in extremely rare circumstances). For practical purposes, the absolute refractive index of air is 1 (it is actually 1.00029 at a temperature of 273 K and a pressure of 101 kPa).

**Practical skills**

Try repeating the refraction experiment, but this time send the ray into the block through the curved side, aimed at the centre of the flat face. This ensures that the ray enters the block along the normal so there is no refraction at this first interface.

Is there always a refracted ray this time? Is there always a reflected ray in the block?

What do you notice about *i, r,* sin *i* and sin *r* this time?

Draw a graph of sin *i* against sin *r*. Compare it with the previous graph. What do you notice about the gradient?

When the direction of the light in the original refraction experiment is reversed, the ray emerges along the same direction as the original incident ray. We say that **light is reversible**. For your experiment $\dfrac{\sin \theta_1}{\sin \theta_2} = {_1n_2}$ and $\dfrac{\sin \theta_2}{\sin \theta_1} = {_2n_1}$. This means that $_1n_2 = \dfrac{1}{_2n_1}$. (Try substituting the absolute refractive indices into this expression and you will see this must be true.) For example, if the refractive index going from air to glass is 1.5, then the refractive index going from glass to air is $\dfrac{1}{1.5} = \dfrac{1}{\left(\dfrac{3}{2}\right)} = \dfrac{2}{3} = 0.67$.

## Worked example: Refractive index and Snell's law

6.  The speed of light in a vacuum is $3.00 \times 10^8$ m s$^{-1}$. Calculate the speed of light in glass of refractive index 1.61.

*Solution*

$n = \dfrac{c}{v}$ so $v = \dfrac{c}{n}$. $v = \dfrac{3.00 \times 10^8}{1.61} = 1.86 \times 10^8$ m s$^{-1}$

7.  A light ray travels from water of refractive index 1.33 to glass of refractive index 1.50. The angle of incidence in the water is 35°.

Calculate the angle of refraction in the glass.

*Solution*

Using the notation a for air, g for glass and w for water, $_a n_w = 1.33$ and $_a n_g = 1.50$.

So $_a n_w \sin i = _a n_g \sin r$ (you can show this using Snell's law, and it's worth remembering).

Therefore $\sin r = \dfrac{_a n_w}{_a n_g} \sin i = \dfrac{1.33}{1.50} \times \sin 35 = 0.509$ leading to $r = 31°$.

## Question

4   A water wave of frequency 6.0 Hz travels at a speed of 0.18 m s$^{-1}$ in the deep region of a ripple tank. After passing into a shallow region, the wave speed decreases to 0.12 m s$^{-1}$.

   a)   Calculate the wavelength of the waves in **i)** the deep region, and **ii)** the shallow region.

   b)   The wave is incident on the boundary between the deep and shallow regions at an angle of 25° to the boundary.

   Determine the refracted angle when the wave enters the shallow region.

5   The speed of light in a vacuum is $3.00 \times 10^8$ m s$^{-1}$. The refractive index of water is 1.33 and the refractive index of glass is 1.52.

   a)   Calculate the speed of light in **i)** the water, and **ii)** the glass.

   b)   A light ray is incident on the boundary between the water and the glass at an angle of 50° in the water. Calculate the refracted angle in the glass.

## Total internal reflection

When you repeated the refraction experiment sending the light into the block through the curved side, you probably found that you could not obtain a final refracted ray for all angles of incidence. Above a certain angle of incidence (40–45° for a glass or plastic block) only a reflected ray occurs (figure 20).

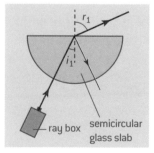 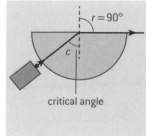 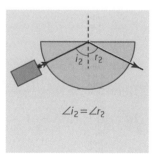

**Figure 20.** Total internal reflection

When light goes from a medium of high *optic density* (with a large refractive index) to one of low optic density (small $n$), such as from glass to air, then the angle of refraction is always larger than the angle of incidence—the direction of the ray moves away from the normal. There is always a ray reflected into the medium with larger $n$. This ray is weak up until the point at which the refracted ray direction is along the boundary (figure 20). For further increases in $i$, no refracted ray is possible and the reflected ray is now strong. This *total internal reflection* occurs for all angles of incidence greater than the *critical angle c*, which is the angle of incidence for which the angle of refraction is 90°. Algebraically:

$$\frac{\sin \theta_1}{\sin \theta_2} = {}_1n_2 \text{ or } \frac{\sin c}{\sin 90} = {}_1n_2 \text{ but } \sin 90 = 1 \text{ so } \sin c = {}_1n_2.$$

With a glass block, ${}_1n_2$ is ${}_{glass}n_{air}$ for total internal reflection. We normally calculate ${}_{air}n_{glass}$, so in terms of the absolute refractive index $n$ of a medium, the critical angle is given by $\sin c = \dfrac{1}{n}$.

### Maths skills

$\sin^{-1} x$ means "the angle whose sine is $x$ is...". To calculate $x$, enter the sine on your calculator—ensuring that it is set to work in degrees, **not** radians—and press $\sin^{-1}$ (or ASIN). Something similar applies for $\cos^{-1}$ and $\tan^{-1}$ (the alternative buttons may be ACOS and ATAN).

### Worked example: Calculating the critical angle

8.  Calculate the critical angle for glass of refractive index 1.52.

*Solution*

$$\sin c = \frac{1}{n} = \frac{1}{1.52} = 0.658$$
$$c = \sin^{-1}(0.658) = 41°$$

9.  The refractive index of water is 1.33 and the refractive index of glass is 1.52. Deduce the critical angle when light goes from water to glass.

*Solution*

These values of $n$ are for air–water and air–glass. First we need to evaluate ${}_{water}n_{glass}$.

$${}_{water}n_{glass} = {}_{water}n_{air} \times {}_{air}n_{glass} \text{ and } {}_{water}n_{air} = \frac{1}{{}_{air}n_{water}}$$

(Treat a new refractive index that you don't know as a multiplication of ones that you do, as here. The "air" subscripts "cancel out" in the multiplication.)

Therefore ${}_{water}n_{glass} = \dfrac{1}{1.33} \times 1.52 = 1.14$ and $\sin c = \dfrac{1}{1.14} = 0.875$.

This leads to $c = 61°$.

## Question

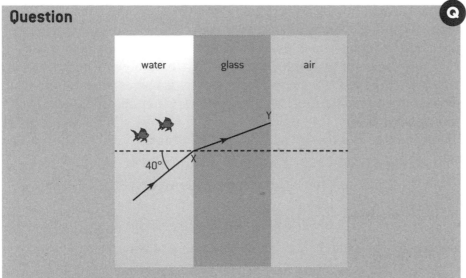

6 A light ray is incident at an angle of 40° on a water–glass boundary at point X. The glass boundaries are parallel and on the other side of the glass is air. The refractive index of water is $\frac{4}{3}$ and the refractive index of glass is $\frac{3}{2}$.

   **a)** Calculate **i)** the angle of refraction at X and **ii)** the angle of incidence at Y.

   **b)** Determine the path of the ray as it leaves Y.

7 A glass window of refractive index 1.51 is covered with a sheet of transparent film of refractive index 1.40. Light is incident at an angle of 45° on the film.

Deduce the subsequent path of the light.

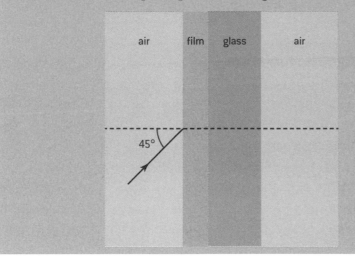

## Optic fibres

The phenomenon of total internal reflection is used in optic fibres to transmit light (figure 21). The fibre is very thin, and the diameter is much exaggerated in the diagram.

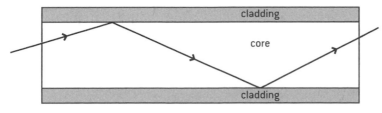

**Figure 21.** Total internal reflection in an optic fibre

The core is surrounded by a cladding with a refractive index less than that of the core. This is precisely the condition for total internal reflection to occur. The rays are incident on the wall of the core at a very large angle of incidence, much greater than the critical angle, and are transmitted along the core. Some energy is lost as the light travels through the fibre, but this loss is tiny: if sea water absorbed light to the same extent as a typical optic fibre, the sea floor would be visible at the deepest part of the ocean (about 10 km down).

Optic fibres are used in telecommunications because:

- they are immune to electromagnetic interference, unlike electrical cables
- they have much lower energy losses than metal wires
- many signals can be carried on optic fibre telephone networks
- their small size means they can be bundled, again increasing network capacity.

In addition, optic fibres are used for visual examinations in places where it is difficult or unsafe for humans to go: in medical investigations inside patients, or for inspecting drains or the interior of nuclear reactors.

**DP link**

You will take the study of optics further if you study **Option C Imaging**.

### DP ready    Nature of science

**Light pipes**

Optic fibres are now essential for local and global communication. The electromagnetic radiation used is generally infra-red rather than visible light. The use of optic fibres has enabled significant progress in digital technology, as much voice traffic is captured in analogue form and then converted to a digital format.

## 4.3 Electromagnetic radiation

We have discussed the properties of light in some detail. However, visible light is just a small segment of the spectrum of **electromagnetic radiation.** All the waves within this group share similar properties:

- They are considered to be transverse waves.
- They result from the motion of accelerated charged particles or changes in energy of charged particles.
- Their most important property is their frequency; the wavelength of an electromagnetic wave changes when the wave enters a different medium.
- They can travel through a vacuum.
- They all travel at the same speed in a vacuum; this speed, called the speed of light, is $3.0 \times 10^8$ m s$^{-1}$ (as we saw earlier, their speed in air is approximately the same).
- They consist of an electric field at right angles to a magnetic field (figure 22).
- When the electric field is at its maximum strength, the magnetic field is also at its maximum.

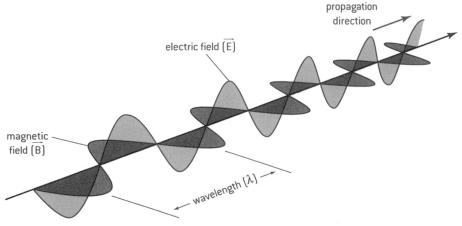

**Figure 22.** An electromagnetic wave

**DP link**

It became clear during the early part of the 20th century that the behaviour of light cannot be explained in terms solely of waves or solely of particles. We now model light as a particle called a **photon** that behaves in ways like both a wave and a particle. You will learn about photons in **12.1 The interaction of matter with radiation**.

The complete range of electromagnetic waves is known as the **electromagnetic spectrum**. Table 1 lists all the classes into which this spectrum is usually divided, with their characteristics and some typical uses.

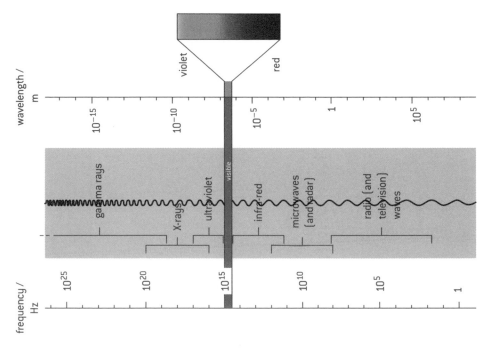

**Table 1.** The electromagnetic spectrum

**Gamma rays** are produced when radioactive nuclei decay or lose energy. Gamma rays have many medical applications, both for imaging inside the body and for treating cancer. They kill bacteria and are used to sterilize food and medical instruments once these items have been sealed into their packaging.

**X-rays** are produced when electrons lose energy rapidly from high speed. As well as medical and dental imaging, X-rays are used in airport security systems and for investigating the internal structure of metal components for safety checks. X-rays are highly penetrating and dangerous to living tissues as they can alter DNA by ionization. Like gamma radiation, X-rays are used to irradiate and destroy cancers.

**Ultraviolet** radiation (UV) is just beyond the short wavelength end of the visible spectrum. Although humans cannot see UV, some insects and other animals can. Some UV from the Sun penetrates the atmosphere and enough arrives at the Earth's surface to damage skin cells. It is used in sunbeds and also to kill bacteria, for example in water treatment.

**Visible** light is the part of the electromagnetic spectrum that the human eye can detect, with a wavelength from about 390 nm (violet) to 700 nm (red). The eye is most sensitive to green wavelengths.

**Infra-red** (IR) radiation can be sensed by the skin, for example when it is emitted by an oven hotplate. It is used for cooking and grilling. A very hot object ("white hot") emits a broad spectrum of radiation from UV through visible to IR.

**Microwave** radiation is part of the radio spectrum. Its uses include transmissions to and from orbiting satellites, and by mobile phones. Also, in microwave ovens, molecules of water, sugars and fats absorb the radiation and heat up. The energy then spreads through the food through thermal transfer. High-power microwaves can damage living tissues through this mechanism.

**Radio waves** have the longest wavelength in the spectrum. With their long wavelengths, some radio waves can diffract around hills and large buildings, so they are used for communication. The radio spectrum includes: long and short waves (wavelengths of kilometres and tens of metres, respectively), vhf (very high frequency; wavelength about 1 m) and uhf (ultra high frequency; wavelength a few cm).

## 4.4 Physics of sound

For most of human history, sound was our principal means of communication. Even today we place great reliance on contacting others using our voices, even though many other methods of communication are available to us.

Sounds travel through solids, liquids and gases. We shall be dealing mostly with sound waves travelling through air. However, much of the physics here also applies to sound waves in liquids and solids. A vacuum prevents the passage of sound; beyond our atmosphere there is only absolute silence.

### Production of sound

Sound is produced when objects vibrate, for example loudspeakers, musical instruments and the human larynx. The vibrations are transmitted through a gas as compressions (high-pressure regions) and rarefactions (low-pressure regions). Figure 23(a) represents these pressure waves moving away from a smartphone, and figure 23(b) shows how the compressions and rarefactions relate to the displacement and pressure variation. The greatest pressure deviations from atmospheric pressure occur where the displacement is zero. Normal atmospheric pressure (the average pressure) is where the particles are displaced by the greatest amount.

⊗ **Internal link**

Look back at **Longitudinal waves** in **4.1 Waves in theory** for a description in the context of the Slinky spring.

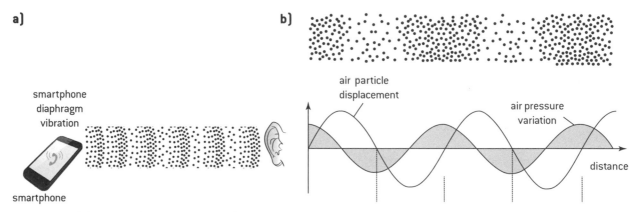

**Figure 23.** (a) Sound (pressure waves) travelling in air; (b) graphs of displacement against distance and pressure variation against distance

---

**DP ready    Nature of science**

The motion of the surfaces of many musical instruments (percussion and strings) is very complex. The photo of a guitar string in figure 24 shows the distortion of the string as it vibrates very quickly. These vibrations are transferred to the air and travel to listeners.

**Figure 24.** The motion of a guitar string captured in a very short period of time

To sum up the properties of sound waves:

- they require a medium (solid, liquid or gas); unlike light, they cannot travel through a vacuum
- they are longitudinal
- they can be reflected and refracted
- they can be diffracted around corners and through gaps.

## Practical skills: Sound needs a medium

In this demonstration, an electric bell connected to a power supply sounds inside a glass jar. As the air is removed from the jar, the sound of the bell outside the jar becomes fainter even though the striker can still be seen hitting the bell. The rubber bands reduce but do not completely eliminate the sound reaching the outside.

## Human hearing

Only a small part of the wide frequency range of sound can be detected by a human ear, from about 20 Hz (a low note on a drum) to 20 kHz (a very high-pitched whistle). This varies greatly between individuals depending on age and other factors; the ability to hear high frequencies decreases with age, and exposure to high sound intensity (loud music, machinery and so on) can damage hearing and restrict the range of frequencies that can be heard.

Sounds with frequencies above 20 kHz are known as **ultrasound.**

## Speed of sound

The wave equation applies to sound just as to other waves and so the speed of sound $c = f\lambda$ as usual. From table 2, showing the speeds of sound in various media, it can be seen that the speeds in gases are low, because the material is not dense (the atoms or molecules are relatively far apart), whereas the speed of sound in solids is usually high as the materials are dense.

**Table 2.** Speed of sound in various media

| Material | Speed of sound (m s$^{-1}$) | Material | Speed of sound (m s$^{-1}$) |
|---|---|---|---|
| air (at 293 K) | 343 | glass | 5640 |
| air (at 273 K) | 331 | copper | 3560 |
| hydrogen (at 273 K) | 1290 | brass | 4700 |
| water | 1490 | lead | 1322 |
| sea water | 1530 | iron | 5130 |

The frequency of a sound is equal to the oscillation frequency of the vibrating object making the sound. Knowing the frequency and the speed enables us to deduce the wavelength of a sound.

## Reflection of sound: Echoes

Just like light, sound can be reflected. The effect is so common that we have additional names for it: echoes, reverberation and so on. When designing concert venues and theatres, architects take great care with the shape of the hall so that reverberation times are appropriate for the type of entertainment that will take place in the hall. Figure 25 shows a concert hall in the UK. The large chambers behind and above the stage area have doors that can be opened and closed to modify the reverberation in the hall for different events.

**Figure 25.** Symphony Hall in Birmingham, UK

Practical uses of echoes include:

- parking sensors in cars, which emit ultrasound and use the time between emission and reception of the sound to determine the distance from sensor to obstacle

- echo-sounders in boats, which send pulses of sound downwards through the water and calculate the distance between the seabed and the sensor below the boat from the time taken for the echo to return; linked to a computer, the sensor can display a map of the local seabed.

### Practical skills: Determining the speed of sound from an echo

Stand about 100 m away from a vertical, straight wall. Measure the distance $d$ to the wall. Clap your hands and listen for the return echo. As you hear the returned sound, clap your hands again and establish a rhythm where you clap exactly as the sound returns. Get a partner to time how long it takes for, say, 10 claps and find the frequency $f$ of your clapping. This will give you the time $t$ between claps. This is the time taken for the sound to travel to the wall and back, in other words $2d$.

So the speed of sound is $\frac{2d}{t}$.

Why would it be poor science to measure the distance to the wall using an ultrasound range finder in this experiment?

## Practical skills: Determining the speed of sound in a solid

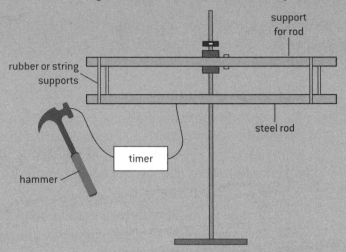

As table 2 shows, the speed of sound in solids is about ten times faster than that in gases. This makes a simple determination of the speed difficult. Here is one way to do it.

Suspend a steel rod horizontally using rubber or string supports. Connect a hammer to an electronic timer so that the timer times only when the hammer is in contact with the rod. Strike the end of the rod with the hammer. When you do this the wave set up by the hammer travels the full length of the rod and back again. The hammer is in contact with the rod while the wave is travelling. But the return of the sound wave to its starting point throws the hammer away from the rod and the electrical contact is broken. So if you know the length of the rod then dividing the distance travelled (there and back) by the time for which the rod and hammer are in contact will give the speed of sound in the rod.

Think carefully about the errors that are likely to arise in this experiment and how you might eliminate them.

## Worked example: Speed of sound and speed of light

10. A girl is 1.5 km from a lightning strike. Calculate the time delay between seeing the strike and hearing the thunder. The speed of sound in air is 340 m s$^{-1}$ and the speed of light is $3.0 \times 10^8$ m s$^{-1}$.

*Solution*

It is possible to calculate the times separately using

$$\text{travel time} = \frac{\text{distance travelled}}{\text{speed of wave}}$$

and then subtract, but a more elegant way is to use algebra:

difference in travel time

$$= \text{distance travelled} \times \left( \frac{1}{\text{speed of sound}} - \frac{1}{\text{speed of light}} \right)$$

$$= \text{distance travelled} \times \left( \frac{\text{speed of light} - \text{speed of sound}}{\text{speed of sound} \times \text{speed of light}} \right)$$

The speed of sound is so much less than the speed of light that the numerator is simply **speed of light** which then cancels with the same term in the denominator, so that, approximately

$$\text{difference in travel time} = \frac{(\text{distance travelled})}{(\text{speed of sound})}$$

$$= \frac{1500}{340}$$

$$= 4.4 \text{ m s}^{-1}$$

8  Radio waves reflected from the surface of the Moon can be detected 2.5 s after they were transmitted from Earth. The speed of electromagnetic waves in a vacuum is $3.0 \times 10^8$ m s$^{-1}$. Calculate the distance between the Earth and the Moon.

9  A longitudinal sound wave travels in air with a speed of 340 m s$^{-1}$. The wavelength of the sound is 1 cm. Predict whether this wave can be detected by a human ear.

### Refraction of sound

Table 2 shows that the speed of sound in cool air is slower than the speed in warm air. This can be expected as the average speed of the air particles is less for cooler air. It is reasonable to expect the particles to take more time to transfer the energy in the wave from one particle to another.

This change in speed with temperature helps to explain why distant sounds appear louder at night. Figure 26 shows the behaviour of sound rays in daylight and at night.

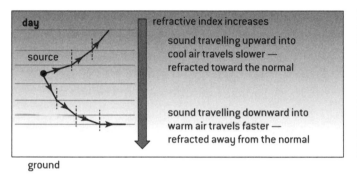 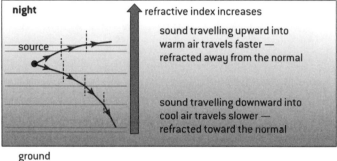

**Figure 26.** Sound refraction in air during the day and at night

- During the day the air near the ground is warm, at a height it is cool. For sound travelling downward from a source, like light rays moving from glass to air, the rays bend away from the normal as they enter a faster medium and so bend upwards as they travel.

- At night the temperature effect reverses. The ground and the nearby air cool rapidly and the air at height is now warmer. Sound rays travelling upwards bend downwards, increasing the number of rays reaching an observer. The sound is louder and seems closer.

This change in refractive index near the ground also explains the **mirage** effect in optics (figure 27). This is an optical illusion often resembling a pool of water: the changes in the refractive index of air lead to rays entering the observer's eye from below rather than above. The brain interprets the image as a reflection in water.

**Figure 27.** A mirage on a hot road looks like a reflecting layer of water

---

**Musical terms**

Musicians have developed terms to describe sound that sometimes differ from those used by physicists to describe the same things. Table 3 compares musical terms with the terms we have used in this chapter.

**Table 3.** Musical and physical terms

| Musical term | Physical term | Meaning | Comparison |
|---|---|---|---|
| pitch | frequency | oscillations per second | high pitch ≡ high frequency |
| loudness | amplitude | maximum displacement of particles in wave | loud sound ≡ large amplitude |

Another term used by musicians is **timbre**: the quality of the sound of an instrument. In scientific terms, the characteristic sounds of musical instrument families relate to how the sounds are produced (stringed instruments: stretched strings; brass instruments: oscillating columns of air; percussion: vibrating wooden or metal plates). Most musical instruments produce a combination of notes with different relative strengths. Differences in the way that the instruments begins to produce its sound, too, are important in our recognition of the instrument.

## Chapter summary

Make sure that you have a working knowledge of the following concepts and definitions:

- ☐ Waves can be mechanical or electromagnetic.
- ☐ Waves can be transverse or longitudinal.
- ☐ Terms to describe the properties of waves include:
    - ○ displacement and amplitude
    - ○ troughs and crests
    - ○ wavelength, frequency and time period.
- ☐ Energy is transferred during wave motion, but the medium is not permanently displaced.
- ☐ Waves can be visualized using wavefronts and rays.
- ☐ The wave speed $c$ of a wave is given by $c = f\lambda$.
- ☐ Graphs of displacement–distance and displacement–time can be used to describe waves.
- ☐ The intensity of radiation varies according to an inverse square law.
- ☐ Diffraction occurs when waves travel past an obstacle or a gap.
- ☐ Light is reversible.
- ☐ Light can be reflected.
- ☐ Images can be real or virtual.
- ☐ Effects observed when light moves from one medium to another include:
    - ○ refraction
    - ○ total internal reflection
    - ○ critical angle.
- ☐ The degree of refraction observed depends on the relative refractive index of the two media involved; this is Snell's law.
- ☐ Optic fibres are an application of total internal reflection used for communication purposes.
- ☐ Light is part of the electromagnetic spectrum.
- ☐ Other parts of this spectrum include:
    - ○ gamma rays
    - ○ X-rays
    - ○ ultraviolet
    - ○ infra-red
    - ○ microwaves
    - ○ radio waves.
- ☐ Light is now thought to consist of photons that possess some wave properties and some particle properties.
- ☐ Sound waves in a gas are longitudinal and consist of regions of compression and rarefaction.
- ☐ Sound waves share some properties with light waves including:
    - ○ echoes as reflection
    - ○ temperature effects similar to the light mirage.
- ☐ Ultrasound is sound at too high a frequency for humans to hear.

## Additional questions

1. **a)** Distinguish between transverse and longitudinal waves.
   **b)** For each of the following, state whether the wave is longitudinal or transverse:
   **i)** radio; **ii)** microwave; **iii)** sound wave in air.

**2.** A hammer strikes the end of a horizontal steel rod of length 1.4 m. The time taken for the sound to travel to the far end of the rod and to return is measured electronically. Four measurements of this time are made and are: 0.43 ms, 0.51 ms, 0.53 ms and 0.45 ms.

Estimate the speed of sound in the steel.

**3.** An ultrasound wave is transmitted into a medium. The graph shows the variation of displacement of the particles in the medium with distance along the medium for one instant. The speed of the wave is 1.2 km s$^{-1}$.

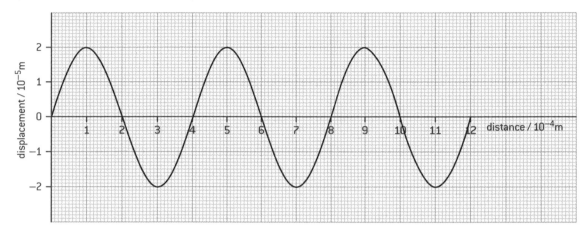

**a)** State the amplitude of the wave.

**b)** Determine the frequency of the oscillation.

**c)** The wave is reflected by an object in the medium and detected. There is a time delay of 95 μs between the wave being transmitted and received. Calculate the depth of the object below the surface.

**4.** A boat is moored in a harbour that has one small entrance. Water waves are moving towards the harbour entrance as shown.

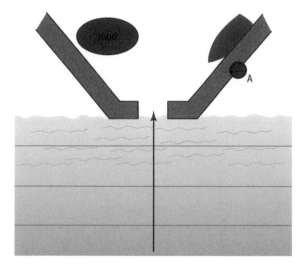

**a)** Describe what will happen to the waves when they arrive at point A on the outer harbour wall.

**b)** Outline why the boat is disturbed by waves even though it is well away from the entrance.

**c)** There is a submerged mudbank in the harbour. When waves pass over this, their wave speed decreases. Explain what happens to the waves as they cross the mudbank.

**5.** A sound wave is emitted from a point source with an output power of 1.5 mW.

Calculate the intensity at a distance from the source of **a)** 2.0 m; **b)** 6.0 m.

6. The refractive index of diamond is 2.42. The refractive index of water is 1.33.

   a) Calculate the critical angle for a diamond **(i)** when it is in the air; **(ii)** when it is submerged in water.

   b) Suggest the effect that submerging the diamond in water is likely to have on the appearance of the diamond.

7. A ray of light, incident on a cube of glass, enters the centre of one face as shown. The ray meets another face at 50°. The critical angle at the glass–air boundary is 45°.

   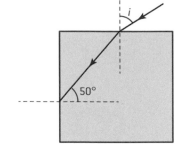

   a) Calculate the refractive index of the glass.

   b) Determine the angle of incidence $i$.

   c) Copy the diagram and use the diagram to predict the path of the ray until it leaves the glass.

8. The core of an optic fibre has a refractive index of 1.51 and a cladding of refractive index 1.36.

   a) Determine the critical angle at the core–cladding boundary.

   b) Sketch the path of a ray that is incident on the core–cladding boundary at an angle greater than the critical angle.

   c) Explain the advantage of using cladding.

9. A ship sounds one blast of its horn. The echo from a cliff that is 1070 m away is heard 6.5 s later. Calculate the speed of sound in air.

10. A student is listening to music played through a loudspeaker.

    a) Explain how the sound travels from the loudspeaker to the student's ears.

    b) A note in the music has a frequency of 500 Hz. The speed of sound in air is 340 m s$^{-1}$. Calculate the wavelength of the sound wave in the air.

    c) Calculate how many times the frequency of the musical note can be doubled before it becomes inaudible to the student, who has normal hearing.

11. A sound has a frequency of 2.0 kHz. The speed of sound in air is 340 m s$^{-1}$.

    a) Calculate the wavelength of the sound in air.

    b) The sound is transmitted into water and the wavelength increases by 0.53 m. Determine the speed of the sound in water.

    c) A ship is sailing above a shoal of fish. A transmitter on the ship sends a pulse of ultrasound vertically down to the fish. The ultrasound signal returns in 0.12 s.

       i) Use your result from **(b)** to determine the distance between the shoal of fish and the ship.

       ii) State **one** other use for ultrasound.

12. a) Complete the diagram by adding the remaining parts of the electromagnetic spectrum.

    |  | X-rays |  |  | infra-red | microwaves |  |
    |---|---|---|---|---|---|---|
    |  |  |  |  |  |  |  |

    b) An X-ray photograph shows a patient's broken ankle, repaired with metal pins to hold it together while the new bone forms. Explain why the bones and the metal pins appear white in the photograph.

    c) Draw a line to join each electromagnetic wave to an application or effect.

    | gamma rays |
    |---|
    | microwaves |
    | ultraviolet |
    | visible light |

    | photosynthesis |
    |---|
    | suntan |
    | cooking |
    | cancer treatment |

# 5 Atomic physics and radioactivity

> **"** It was quite the most incredible event that has ever happened to me in my life. It was almost as incredible as if you fired a 15-inch shell at a piece of tissue paper and it came back and hit you. **"**
>
> **Ernest Rutherford, discussing the result of the Geiger–Marsden experiment (1938)**

## Chapter context

So far in this book, **atoms** have been modelled as small particles with mass but no other features. Now we shall look at the **interior** of the atom and how it can change. This area of physics stems largely from the end of the 19th century onwards; earlier attempts to explain the nature of atoms were either speculative or philosophical.

## Learning objectives

In this chapter you will learn about:

→ the **composition** of the **atom** and its **nucleus**

→ **Rutherford's model** for the atom and the evidence for it

→ **nuclear changes** during **radioactive decay**

→ ways to detect nuclear **radiation**

→ **background** radiation and its causes

→ safe use of radioactive material

→ the concept of **half-life** and its determination

→ the use of **error bars**

## Key terms introduced

→ proton number and nucleon number

→ isotopes

→ nuclide

→ proton, neutron, neutrino, electron and positron

→ ionization

→ alpha, beta and gamma radiation

→ background count

→ activity and half-life

→ becquerel

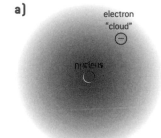

a)

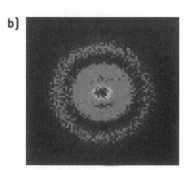

b)

**Figure 1.** (a) Electron probability cloud around a ground-state hydrogen atom; (b) photograph of the charge distribution around an excited-state hydrogen atom

## 5.1 Inside the atom

On the basis of much work by many scientists, the atom is now thought to consist of a small, dense nucleus surrounded by atomic *electrons*. Textbooks often show atoms with electrons as though they are in orbit in order to illustrate the arrangement of the electrons outside the nucleus. This arrangement is important for the chemistry of the element. However, you should not regard electrons as small, definite points (as we did implicitly in *2 Electric charge at work*). The present model treats electrons as probability clouds, which indicate the probability of where the electron is likely to be found (figure 1(a)). Figure 1(b) shows one of the first images of the charge distribution in a single hydrogen atom. It indicates that the electron does not orbit the nucleus.

## Key term

A **proton** has a positive charge of the same magnitude as that of an electron.

A **neutron** has a similar mass to that of the proton but no charge, and is therefore electrically neutral (hydrogen is the only element without a neutron in the nucleus).

The mass of a proton is $1.673 \times 10^{-27}$ kg.

The mass of a neutron is $1.674 \times 10^{-27}$ kg.

For comparison, the mass of the much smaller, negatively charged **electron** is

$9.110 \times 10^{-31}$ kg, about $\dfrac{1}{1800}$ of that of the proton.

### DP ready  Approaches to learning

### Referencing work

The image of the excited hydrogen atom was produced in 2013 by an international team of scientists. It was originally published in a scientific journal called *Physical Review Letters.* Here is the complete bibliographical reference, which allows anyone else to find the publication unambiguously:

AS Stodolna, A Rouzée, F Lépine, S Cohen, F Robicheaux, A Gijsbertsen, JH Jungmann, C Bordas, MJJ Vrakking. Hydrogen Atoms under Magnification: Direct Observation of the Nodal Structure of Stark States. *Phys. Rev. Lett.* 2013, *Vol. 110*, pp. 213001–6.

The basic information required is: Authors' surnames + initials / Article title / Journal title, volume, part number and page numbers / Date of publication.

When you are referencing work in your Diploma Programme, you should always give full credit to any work by other people that you use. This also enables your readers to learn more and to check that you have drawn the correct conclusions from other work.

Inside the nucleus are two varieties of nucleon: the positive *proton* and the neutral *neutron*.

Table 1 shows the structure of five atoms common on Earth: hydrogen, helium, carbon, oxygen and iron.

**Table 1.** Five common atoms and their composition

| Element | | Proton number | Nucleon number | Symbol |
|---|---|---|---|---|
| hydrogen-1 | 1 electron, 1 proton, 0 neutrons | 1 | 1 | $^{1}_{1}\text{H}$ |
| helium-4 | 2 electrons, 2 protons, 2 neutrons | 2 | 4 | $^{4}_{2}\text{He}$ |
| carbon-12 | 6 electrons, 6 protons, 6 neutrons | 6 | 12 | $^{12}_{6}\text{C}$ |
| oxygen-16 | 8 electrons, 8 protons, 8 neutrons | 8 | 16 | $^{16}_{8}\text{O}$ |
| iron-56 | 26 electrons, 26 protons, 30 neutrons | 26 | 56 | $^{56}_{26}\text{Fe}$ |

nucleon number

$^{A}_{Z}\text{X}$ ← chemical symbol for element

proton number

### DP ready  Nature of science

### Describing an atom

The notation used for atoms is $^{A}_{Z}\text{X}$, where X is the chemical symbol of the element, *A* is the *nucleon number* (sometimes called the *mass number*)—the number of nucleons in the element—and *Z* is the *proton number* (sometimes called the *atomic number*). This symbol also tells you the number of neutrons in the nucleus ($A - Z$) and the number of electrons in a neutral (un-ionized) atom (*Z*). The nitrogen in the atmosphere is either $^{14}_{7}\text{N}$ (99.6% of the nitrogen) or $^{15}_{7}\text{N}$ (the rest): all nitrogen atoms have seven protons and seven electrons, with either seven or eight neutrons.

You will also see atoms written as helium-4 (He-4). This notation gives only the nucleon number, but is useful when distinguishing between *isotopes* of an element.

The isotopes of a single element all have the same chemical behaviour but different physical characteristics, because the chemistry of an atom is determined by the numbers and configuration of the electrons, but its physical properties depend on its mass. The proton numbers of two isotopes of the same element are the same by definition, but the nucleon numbers will be different. Each different atom is known as a *nuclide*.

Sometimes there is more than one stable isotope of an element. For example, iron (Fe) has four naturally occurring isotopes, Fe-54, Fe-56, Fe-57 and Fe-58. The proton number of these isotopes is always 26.

### DP ready    Approaches to learning

The terms nuclide and isotope are often confused. Strictly the word isotope should be used in the plural or with reference to another nuclide: "The nuclide $^{14}_{6}C$ is an isotope of $^{12}_{6}C$."

### Key term

The **nucleon number** or **mass number** $A$ is the total number of protons and neutrons in the element.

The **proton number** or **atomic number** $Z$ is the number of protons in the nucleus.

**Isotopes** are versions of the same element with different numbers of neutrons in the nucleus.

A **nuclide** is an individual species of atom, such as a particular isotope.

### Worked example: Using atomic symbols

1.  A form of carbon has the symbol $^{14}_{6}C$. Deduce the number of protons, neutrons and electrons in one uncharged atom of this nuclide.

*Solution*

There are six protons in the nucleus and there must be six electrons in the electron shells for the atom to be neutral. The total number of nucleons in the nucleus is 14 so carbon-14 must have $(14 - 6) = 8$ neutrons.

2.  Helium-3 is an isotope of helium. Deduce the atomic symbol for this isotope.

*Solution*

Helium is the second element in the periodic table, so it has two protons, $Z = 2$. Therefore, there must be one neutron. The symbol is $^{3}_{2}He$.

## Rutherford's model of the atom

In 1909 Rutherford was advising his research students, Johannes Geiger and Ernest Marsden, who were investigating the deflection of alpha particles ($^{4}_{2}He$ nuclei) by gold atoms in a very thin foil. The alpha particles were detected when they collided with a zinc sulfide screen and caused fluorescence. Rutherford suggested to Geiger and Marsden that they should move their detector to check whether any alpha particles were deflected through very large angles. To the astonishment of all three scientists, they found that about 1 in 8000 of the alpha particles were "reflected" by the thin foil which was only a few atoms thick (figure 2). This was the moment when Rutherford realized that, in his words, "*..the scattering backwards must be the result of a single collision…[and]…I had the idea of an atom with a minute massive centre, carrying a charge.*"

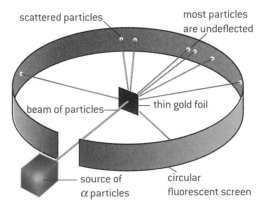

**Figure 2.** The Rutherford–Geiger–Marsden experiment

**Internal link**

The origin of alpha particles is explained in **5.2 Radioactive decay**.

This "reflection" or, more correctly, back-scattering is only possible because the massive part of the atom is positively charged. Rutherford was able to show that the atom has a positive nucleus with a diameter of about $10^{-15}$ m and that the whole atom, including the electrons outside the nucleus, has a diameter of about $10^{-10}$ m. Atoms are essentially empty space.

We now know that the deflection of the alpha particle is caused because there is an electric repulsion force acting between the positively charged alpha particles and the highly positive nuclei of the gold atoms in the foil.

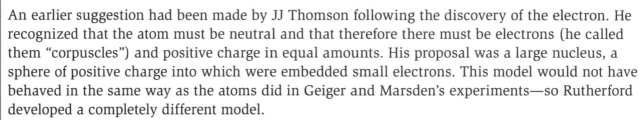

**DP ready** | **Nature of science**

**An earlier model of the atom**

An earlier suggestion had been made by JJ Thomson following the discovery of the electron. He recognized that the atom must be neutral and that therefore there must be electrons (he called them "corpuscles") and positive charge in equal amounts. His proposal was a large nucleus, a sphere of positive charge into which were embedded small electrons. This model would not have behaved in the same way as the atoms did in Geiger and Marsden's experiments—so Rutherford developed a completely different model.

Thomson's model was called the "plum pudding model" as it resembled a British pudding traditionally eaten at Christmas, which has small pieces of fruit embedded in it, like Thomson's proposal for the electrons in the atom!

These sizes are difficult to comprehend. A scale model helps: imagine an atom nucleus scaled up to the size of a soccer ball (22 cm in diameter) and placed in the centre of a soccer pitch. The edge of the scaled-up atom will be about $1.1 \times 10^4$ m or 11 km away from the ball.

At the time of the Geiger–Marsden experiment, the neutron had not been detected. In 1930, Bothe and Becker in Germany found that an unknown particle was emitted from beryllium when it was bombarded by alpha particles. Two years later, Chadwick in the UK showed that this particle was neutral, and it was named the neutron. Over the next years, a whole family of particles was found to be emitted from atoms under various conditions. These findings are summarized in the Standard Model, which is now used to explain the changes that occur in nuclei.

**Internal link**

The Standard Model is introduced in **5.4 The Standard Model** at the end of this chapter.

**Question**

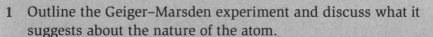

1 Outline the Geiger–Marsden experiment and discuss what it suggests about the nature of the atom.

2 In the Geiger–Marsden experiment some alpha particles are deflected through small angles while a few are deflected through very large angles. Suggest why the Thomson "plum pudding model" of the atom can explain small deflections of the alpha particles, but cannot explain large deflections.

## 5.2 Radioactive decay

Not all nuclei are stable. Some nuclei change spontaneously and randomly to form a new nuclide. As they do so, they emit nuclear radiation in the form of small particles, usually with some electromagnetic radiation in addition. This behaviour is known as **radioactive decay**.

## The random nature of radioactivity

Many events in science are predictable. Apply a force to an object and it will accelerate by an amount predicted by $F = ma$. Radioactivity is unusual in that it is a random event. There is nothing we can do to influence when a particular nucleus will decay. Pressure and temperature have no effect and neither does the presence of other particles.

## A brief history of radiation

The particles and radiation emitted by unstable atoms are highly energetic and can *ionize* the matter through which they pass.

We now know that there are three principal types of radiation that emerge from unstable nuclei: *alpha* (α) and *beta* (β) particles, and *gamma* (γ) radiation. Their properties were observed by scientists about a decade before they were completely identified.

## Alpha particles

Alpha radiation was first observed by Rutherford in 1899. He found that it was strongly ionizing but did not penetrate more than a single sheet of thin paper or foil (figure 3).

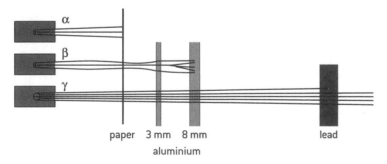

**Figure 3.** Absorption of radiation by different absorbers

In fact, these two observations are essentially the same because alpha radiation is so strongly ionizing (requiring energy) that the alpha particles quickly run out of energy after a succession (usually thousands) of ionization events in the absorbing material. In 1909 Rutherford and his collaborator Royds identified alpha radiation: a **doubly ionized helium atom**, that is, two protons and two neutrons with no electrons. This helium nucleus comes from inside the unstable nucleus.

### Key term

**Ionization** is the removal or addition of electrons from or to atoms. We need only consider the removal process, but the addition of electrons to atoms is important in chemistry.

The energy of particles emitted in radioactive decay is sufficient to knock one or more outer electrons away from a nearby atom. This makes the atom (now called an **ion**) positively charged. A gas that has been ionized can carry current, or if the ionizing radiation passes through living tissue it can damage and destroy the living cells.

One of the reasons why alpha particles are so strongly ionizing is their charge of +2e. They can therefore exert a strong force on nearby electrons. Also, their mass is relatively large and so they travel relatively slowly and spend more time near atoms. This allows more opportunity for the ionizing events to take place.

## Beta particles

Beta radiation is much less ionizing than alpha radiation and is therefore much more penetrating. Typically, a beta particle will penetrate air and liquid but will be stopped by a few centimetres of aluminium or a few millimetres of lead.

Beta particles are electrons that have been emitted by the unstable nucleus when it decays. Beta particles were identified quickly, as in 1900 Becquerel was able to show that the ratio

$$\frac{\text{charge on the beta particle}}{\text{mass of the beta particle}}$$ was exactly the same as for the electron

(which had been identified a few years earlier). This was a very strong suggestion that the beta particle was an electron, but from the interior of the nucleus.

## Gamma radiation

The third emission from the nucleus is not a particle at all but energy from the electromagnetic spectrum. These emissions are highly penetrating and can go through metres of lead and concrete without being absorbed. Conversely, these radiations are not highly ionizing. Paul Villard discovered the radiation in 1900 and it was named "gamma radiation" by Rutherford in 1903 to differentiate it from the alpha and beta that he had already named. However, it was not until 1914 that he was completely satisfied that it was electromagnetic.

## Distinguishing alpha, beta and gamma

Rutherford originally thought that gamma rays were very energetic beta particles. A comparatively simple experiment (figure 4) disproved this: a radioactive source, located in a block that allows a thin beam of radiation (a "collimated beam") to emerge, is placed next to a strong magnetic field. Gamma rays, which are uncharged, are not deflected by the field and travel straight on without deviation.

Alpha particles are positively charged and are therefore deflected like conventional current. In figure 4 they are accelerated upwards and strike the screen above the gamma rays. Use Fleming's left-hand rule or another rule to verify that the deflection of the alpha particles is consistent with a positive charge moving in the magnetic field shown.

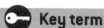
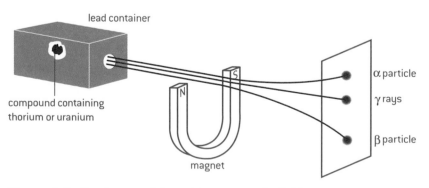

**Figure 4.** Deflecting particles with a magnetic field

Beta particles are electrons. They are deflected much more than alpha particles because they have a very small mass, and because their charge is negative they are deflected in the opposite direction to alpha particles. Again, you can check figure 4 using your preferred direction rule.

**Internal link**

The direction rule for the motor effect is described in **2.3 Magnetism at work.**

**Table 2.** Summary of the radiations and their properties

| Radiation | Constituents | Range | Ionizing ability |
|-----------|--------------|-------|------------------|
| alpha $\alpha$ | helium nucleus $^4_2\text{He}^{2+}$ | few centimetres of air; thin sheet of paper | high |
| beta $\beta$ | electron *from nucleus* $^0_{-1}e^-$ | few centimetres of aluminium | medium |
| gamma $\gamma$ | electromagnetic radiation | metres of concrete and lead | low |

## Detecting radiation

**Cloud chamber:** One detector used in the early days of research into radioactivity was an instrument called the cloud chamber. The form still found in some school laboratories is the diffusion type. A cold material such as dry ice (solid carbon dioxide) is used to cool the base of a small chamber lined with a pad soaked in volatile liquid, for example, alcohol. The evaporation of the liquid and the temperature gradient lead to the formation of a supersaturated air/vapour mixture at the bottom of the chamber. A supersaturated vapour is one in which the vapour is too concentrated for the temperature. When an ionizing particle travels through the chamber, the vapour condenses onto the ions produced by the particle. This condensation path is seen as a track in the chamber (figure 5).

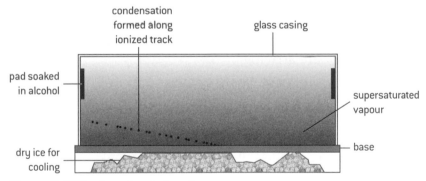

**Figure 5.** Diffusion cloud chamber

Figure 6 shows an event marked by tracks in a cloud chamber. An alpha particle is emitted from a radioactive lead source and is scattered at various points. The cloud chamber, and its successor the bubble chamber (which used superheated liquid hydrogen rather than a supersaturated vapour), were used for many years to record the tracks of particles and to carry out fundamental research into nuclear changes.

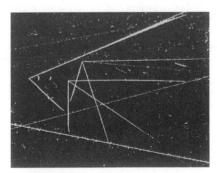

**Figure 6.** Tracks in a cloud chamber

**Geiger–Müller tube:** A convenient device frequently used in school laboratories is the Geiger–Muller tube (figure 7), which gives a count of how many particles enter it rather than an image of a particle track. The method of operation relies on ionization, like many other detection devices.

## Internal link

The discussion of the rapid change in speed of the electron and the formation of an electron cascade links to much of **2 Electric charge at work**.

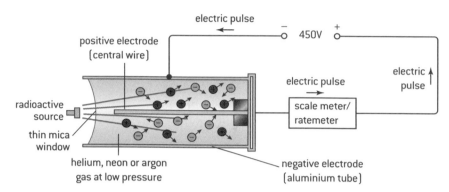

**Figure 7.** The Geiger–Müller tube

A sealed tube contains a mixture of gases at low pressure. The tube also contains a central wire electrode that is positively charged. The outer wall of the cylindrical container is negatively charged so there is an electric field between the centre and outside of the tube. The slightly curved window at the end of the tube nearest the source is made of mica, so thin that incident alpha particles can pass through it.

When ionizing radiation enters the tube it ionizes the gas atoms to form an ion and an electron. Both are subject to an electric force but the electron, with its small mass, is accelerated much more than the ion. It gains energy quickly and ionizes other atoms along its path. Very soon a large number of electrons have formed from the single initial ionization by the incident particle. This is called a **cascade process**. Essentially a spark jumps between the wall and the central electrode and is detected as a current in the electrical circuit. A meter can be used to detect either the arrival of each individual particle, or the rate at which they are arriving.

You may wonder why the spark, once begun, ever stops. Atoms of another gas in the tube, known as a **quenching agent**, slow down the electrons after a very short time and end the discharge in the tube. While the quenching occurs, no further count can be made. Manufacturers of Geiger–Müller tubes work hard to make this down-time as brief as possible.

### Background radiation

## Key term

**Background radiation** is the ionizing radiation always present from existing, mostly natural, sources in the environment other than the source under consideration.

Switch on a Geiger–Müller tube and counter; even with no radioactive source in the laboratory you will observe radioactive counts. This is the **background count**, measuring *background radiation*. There is radioactive material everywhere: in rocks, the air we breathe, the water we drink and in plants and animals. Figure 8 lists some of sources of background radiation and their relative sizes. You will see that a major component is radon gas (Rn-86), which occurs naturally in rocks and can accumulate in buildings that are built over those rocks or from materials derived from them.

## Practical skills: Using radioactive sources

The sources used in schools are always very weak and completely safe for your use, provided that you always follow the advice of your teacher and you observe some basic precautions.

- Use tongs to handle them; never touch the sources with your fingers.
- Never hold the sources near your eyes.
- Never point them at anyone else.
- Keep them in their protective boxes when they are not in use.
- Store them in locked cabinets when the laboratory is not staffed.

The crucial ways to protect people against radiation are to ensure distance from the source, shielding in the form of absorbers between source and experimenter, and minimal time for exposure to the source. These principles underpin the safety precautions for those who work professionally with radiation, such as doctors and nuclear workers.

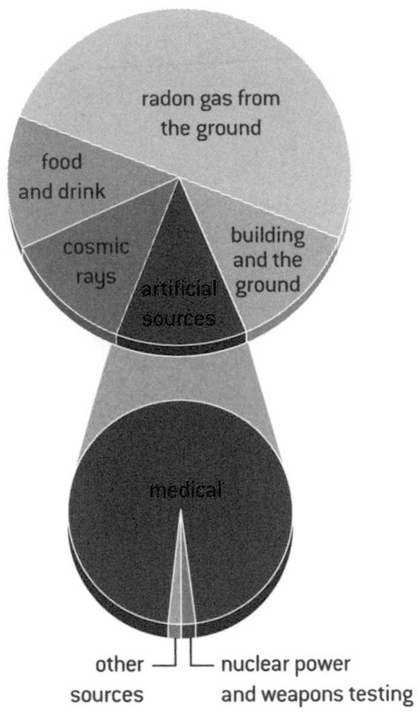

**Figure 8.** Sources of background radiation

### DP ready | Nature of science

#### International issues

The International Atomic Energy Authority is a body committed to providing "…a strong, sustainable and visible global nuclear safety and security framework". It works to protect people and society from the harmful effects of ionizing radiation. Such bodies come under the auspices of the United Nations organization. Over the years many scientists have contributed to international work seeking to spread information about safety for the public and for workers in the nuclear industry.

## Public safety

There is always public concern about the safety of radioactive sources and worry about exposure to radiation. Given the absorption properties of the three types of ionizing radiation:

- alpha particles are readily absorbed so, unless the source of the radiation is swallowed or breathed in, danger from alpha radiation is relatively low
- beta and gamma radiations are more penetrating and can cause radiation damage and long-term alteration to DNA.

Nuclear radiations can be handled safely, and people who work regularly with radiation are closely monitored so that they do not suffer an overdose.

## Decay equations

Having studied the nature and properties of the three types of emitted radiation, we will now look at how and why these radiations are emitted from the nucleus. Nuclear equations are used to show the original nucleus and its products.

**Alpha decay:** an unstable nucleus emits a single particle that is identical to a helium nucleus (in other words, two protons and two neutrons). Heavier elements are most likely to decay by this route.

Typical examples are:

the alpha decay of uranium-238 to thorium-234: $^{238}_{92}\text{U} \rightarrow \, ^{234}_{90}\text{Th} + ^4_2\text{He}$

the alpha decay of radon-198 to polonium-194: $^{198}_{86}\text{Rn} \rightarrow \, ^{194}_{84}\text{Po} + ^4_2\text{He}$

In both decays there are only two products, the alpha particle and a new nucleus, often called the **daughter nucleus**. Look carefully at the values of $A$ and $Z$ in the equations; they balance so that the totals on the right are equal to the values on the left. In addition, energy is released from the nucleus during the decay. This is not usually written in the nuclear reaction, but it has important consequences.

**Internal link**

Conservation of momentum is discussed in **1.4 Momentum and impulse**.

---

### DP ready — Nature of science

#### Mechanics of alpha decay

Any energy released by the nucleus during alpha decay is transferred to kinetic energy in the emitted particles: the alpha particle and the daughter nucleus. If the initial nucleus was stationary (unusual, but let's assume so) then the initial momentum of the system was zero. By conservation of momentum, it must be zero afterwards too, so the momentum of the two products must sum to zero. Use the expression for momentum, $mv$, to show that the ratio of the speed of the daughter $v_{\text{d}}$ and the alpha $v_\alpha$ must be $\dfrac{v_{\text{d}}}{v_\alpha} = \dfrac{m_\alpha}{m_{\text{d}}}$ where $m_{\text{d}}$ and $m_\alpha$ are the masses of the daughter nucleus and the alpha particle respectively.

When there is no other way to gain or lose momentum, this means the direction of velocities $v_{\text{d}}$ and $v_\alpha$ must be opposite. The daughter nucleus and the alpha move apart at 180° between their velocities. Because the energy released in the reaction usually has only one or two possible values, the sharing of energy between alpha and daughter is predictable and the alpha particles all leave only one or two possible track lengths in a cloud chamber.

---

**Beta decay:** In beta-particle emission, the unstable nucleus emits an electron, $^0_{-1}\text{e}$. The electron has a very small mass, so its nucleon number is zero, and its proton number of −1 describes its charge.

Typical examples are:

the beta decay of carbon-14 to form nitrogen-14: $^{14}_6\text{C} \rightarrow \, ^{14}_7\text{N} + ^0_{-1}\text{e} + \bar{v}_e$

the beta decay of iodine-131 to form xenon-131: $^{131}_{53}\text{I} \rightarrow \, ^{131}_{54}\text{Xe} + ^0_{-1}\text{e} + \bar{v}_e$

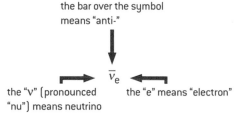

the bar over the symbol means "anti-"

$\bar{v}_e$

the "v" (pronounced "nu") means neutrino       the "e" means "electron"

Again, the numbers balance, so the nucleon number of the parent nucleus is the same as that of the daughter. The proton number increases by 1 as the electron has −1.

More importantly, a new particle appears, written, as you will be expected to write it, as $\bar{v}_e$: the electron antineutrino. This third particle is a hint that something more complicated is happening in the nucleus than has been suggested so far.

## Key term

The **neutrino** was first proposed during the 1930s in response to unexplained discrepancies in the energy of the beta particle emitted in the decay. It is a small massless particle that has momentum.

### DP ready · Nature of science

**Mechanics of beta decay**

When alpha particles are emitted in a decay, only a few different energies are possible for the alpha particle. As explained before, when there is only one possible value for the energy released and only two particles then the speeds for the two particles can only be distributed in one way.

Beta decay is different. The emitted beta particles take a range of energies from zero to a maximum value that is never exceeded. If only two particles were released, this would not be possible. This is how scientists were able to predict the existence of the *neutrino*. It can carry away momentum and energy, and with three particles emitted there is a vast range of possibilities for the distribution of energy and momentum between the daughter nucleus, the beta particle and the antineutrino. So the beta particles can have any energy from zero up to the value corresponding to almost all the energy released by the decaying nucleus.

Another clue that the proton–neutron model is not the final word for the nuclear model is the existence of three more decay modes related to beta decay. In the first of these a **positron** is emitted by the unstable nucleus. The *positron* is the *antiparticle* of the electron. This is known as **beta-plus decay** ($\beta^+$).

## Internal link

You will learn more about subatomic particles in **5.4 The Standard Model**.

## Key term

Antimatter is not science fiction! **Antiparticles** are pairs of particles: they have the same mass but opposite charge, and one is made of antimatter—a **positron** is an antielectron, with a positive charge. When a particle and its antiparticle interact they both disappear ("annihilate") and are replaced by two gamma photons travelling in opposite directions (so that energy and momentum are conserved). The total energy of the photons is equal to the total energy of the antiparticles, including the energy equivalent of their mass.

Every known particle has an antiparticle. However, some particles (like photons) are their own antiparticle.

The nuclide carbon-11 decays by positron emission: $^{11}_{6}C \rightarrow \, ^{11}_{5}B + \, ^{0}_{+1}e + \nu_e$

Note the differences between this equation and that for beta-minus ($\beta^-$) decay:

- the third particle is a neutrino, not an antineutrino (no bar on the top)
- the proton number of the daughter product is one lower than the original nuclide.

The other two decay modes are rare: they involve the capture of either an electron or a positron by an unstable nuclide. There is a symmetry to these four decay modes, with two emissions and two captures of either an electron or its antiparticle.

**Gamma emission:** When an unstable nucleus has decayed through either the alpha or beta route, the daughter product is often left with excess energy. The emission of a gamma ray—a high-energy photon—is an ideal way for the nucleus to return to its **ground state** (unexcited state). There is no change to the daughter other than the loss of energy.

An example is the beta decay of cobalt-60 to nickel-60. The first equation is for the beta decay, the second for the gamma. The asterisk (*) means that the nickel-60 is in an excited state.

$$^{60}_{27}\text{Co} \rightarrow\ ^{60}_{28}\text{Ni}^* +\ ^{0}_{-1}\text{e} + \bar{\nu}_e + \gamma$$

$$^{60}_{28}\text{Ni}^* \rightarrow\ ^{60}_{28}\text{Ni} + \gamma$$

**Worked example: Isotopes and elements**    **WE**

3. The diagram represents three neutral atoms X, Y and Z.

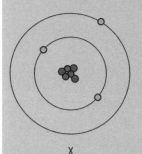 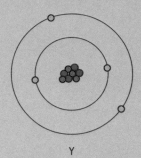 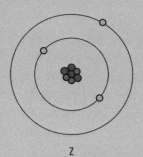

| X | Y | Z |

Deduce which atom is a different element from the other two and which are isotopes of the same element.

*Solution*

The command term "Deduce" requires a high-level explanation; simply giving the atom letter will not score marks.

X and Z have the same number of electrons and therefore, as they are neutral, the same number of protons. The yellow circles in the nucleus are therefore protons. So Y is a different element from the others.

Isotopes are nuclides with the same numbers of protons (and electrons, in a neutral atom) but different numbers of neutrons. X and Z are therefore the isotopes.

**Question**    **Q**

3   Copy and complete the table.

| Particle | Relative mass | Relative charge |
|----------|---------------|-----------------|
| proton   | 1             | +1              |
| neutron  |               |                 |
| electron |               |                 |

4   The diagram shows the nuclei of four atoms, W, X, Y, Z.

W    X    Y    Z

a)   State the nucleon number of each atom.

b)   Identify, with reasons, any atoms that are isotopes.

**Internal link**

To remind yourself of the analogy of the soccer ball and the huge distance from one nucleus to the next, go to the end of **5.1 Inside the atom**.

## 5.3 Half-life

It is now clear why radioactive decay is random and spontaneous; it occurs in the nucleus, which is isolated at the heart of the atom, buffered from other nuclei by atomic electrons. However, the statistics of a large number of particles decaying at random can be analysed using the Poisson distribution.

**The Poisson distribution**

Simeon Denis Poisson (1781–1840) was a French statistician for whom the Poisson distribution was named. This statistical description deals with the occurrence of random events and so describes the chance of an unstable nuclide decaying, as well as:

- the number of calls arriving at a call centre in a given time
- the change in height of a foam with time (the chance of a bubble bursting is random, but with a definite probability)
- the chance of finding a particular species of plant or animal in a biological transect
- the number of mutations on a section of DNA.

One of the joys of science is uncovering the mathematical similarities between apparently dissimilar fields.

Each of the unstable nuclei in a collection of identical atoms has the same **probability of decay** as every other nucleus. This does not depend on the size of the sample. As we shall see, this constant chance of decay per nucleus leads to a simple idea, that of **half-life**. It is best to approach this mathematical idea through a simple experiment that involves another example of the Poisson distribution.

**Practical skills: Modelling radioactive decay**

You will need 100 (or more) dice, each with six sides. The procedure is straightforward.

- Throw all the dice simultaneously and remove every die that lands with a six upwards.
- Note the number removed (decayed).
- Repeat with the remaining dice (showing numbers 1–5). Keep going until there are only a few dice left.
- Calculate the number of dice remaining for each throw and plot it against number of throws on a graph.
- The graph will look something like figure 9.

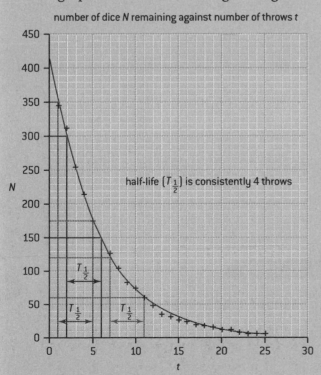

**Figure 9.** Decay curve for sixes thrown with a large number of dice

**Half-life** is the time taken for half the events to occur; for radioactivity, the time for the number of unstable nuclei remaining to halve.

The graph in figure 9 shows a *half-life* behaviour—it always takes the same number of throws to halve the number of dice. Three examples of this are shown on the graph.

Figure 10 shows the curve that is obtained when the number of counts per second (the count rate) registered from a radioactive nuclide by a detector such as a Geiger–Müller tube is plotted with time. Count rate is not always the same as *activity*, measured in *becquerels*, because the detector may not pick up all the decays.

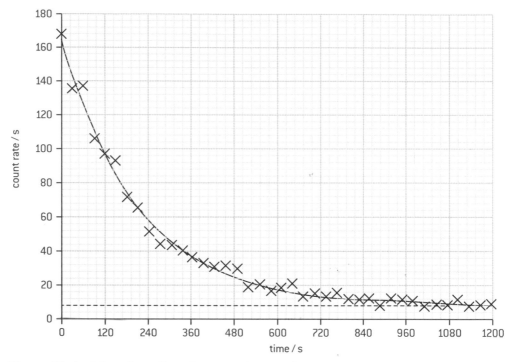

**Figure 10.** Activity of a radioactive sample over time

This graph has several points to notice:

- The value does not drop to zero as the time becomes large; this is because the graph is not corrected for background radiation. You need to take this into account.

- The data points do not lie perfectly on a smooth curve. Poisson's distribution predicts that the error in a data point (the count rate in this case) is equal to plus or minus the square root of the value of the data point ($\pm\sqrt{\text{count rate}}$ here). Thus, when the rate is measured as 100 counts per second, the error will be 10 (that is, $\sqrt{100}$) and we should properly write the value of the data as $(100 \pm 10)$ counts per second.

You should be able to show that the half-life for this decay is about 150 s when you have allowed for the background.

**Key term**

The **becquerel** (Bq) is the SI unit of **activity**. One becquerel is the activity of a sample that decays at a rate of one nucleus per second. Its fundamental unit equivalent is $s^{-1}$.

The unit was named for Henri Becquerel, who shared the Nobel Prize for Physics in 1903 with Marie and Pierre Curie for their discovery of radioactivity.

| DP ready | Approaches to learning |

Some of the many examples of half-life behaviour in science have already been mentioned. It arises when exponential changes are involved in a process. You will meet this in Higher Level Diploma Physics.

Another way to think about this behaviour is that the change is directly proportional to the amount of substance present.

In radioactivity, there is a definite probability that any one nucleus will decay in the next second (called the decay constant, $\lambda$). If there are $N$ nuclei in a sample, then the number that decay in the next second must be $\lambda N$, or more strictly $-\lambda N$ because the number of nuclei is dropping.

It is the same for the foam on a drink. Each bubble has the same chance of disappearing in the next second, so the rate at which bubbles disappear (the height of the foam) must show a half-life behaviour. Try it with a suitable foaming drink!

**Table 3.** Half-lives of some elements

| Unstable element | Half-life | Principal emission |
|---|---|---|
| boron-12 | 20 ms | $\beta^-$ |
| radon-220 | 52 s | $\alpha$ |
| iodine-128 | 25 minutes | $\beta^-$ |
| radon-222 | 3.8 days | $\alpha$ |
| strontium-90 | 28 years | $\beta^-$ |
| radium-226 | 1600 years | $\alpha$ |
| carbon-14 | 5700 years | $\beta^-$ |
| plutonium-239 | $2.4 \times 10^4$ years | $\alpha$ |
| uranium-235 | $7.1 \times 10^8$ years | $\alpha$ |
| uranium-238 | $4.5 \times 10^9$ years | $\beta^-$ |

Table 3 shows the large range of half-lives observed—and there are both longer and shorter half-lives than this. Logically, a long half-life indicates a low activity. For example, compare 1 mol of radon-222 and 1 mol of iodine-128. The samples have the same initial number of radon and iodine atoms. However, after 25 minutes only 0.5 mol of iodine will remain, while the number of radon atoms will barely have altered. The iodine must have a much higher activity than the radon.

**DP ready** — **Nature of science**

Although some decaying elements last a very long time and therefore give many problems in long-term storage, high radioactivity is not necessarily one of the problems. In many cases, the essential problem with these long-lived nuclides is toxicity; they are extremely poisonous to all forms of life.

**Worked example: Determining the half-life** **WE**

4. The graph on the following page shows how the activity of a radioactive nuclide varies with time. Determine the half-life of the nuclide from the graph.

*Solution*

The half-life is the time taken for the activity (or count rate) to fall by half. It takes 7.0 s to fall from 1000 s$^{-1}$ to 500 s$^{-1}$. However, only taking one value is unsafe (notice the error bars on the graph and the uncertainty that they introduce). Always repeat for at least two different values. When the count rate falls from 800 s$^{-1}$ to 400 s$^{-1}$ and from 600 s$^{-1}$ to 300 s$^{-1}$, the half-life is also 7.0 s and this value is confirmed.

## Maths skills: Error bars

A way to represent error on a graph is to use error bars. Look at the graph carefully and you will see small capped lines ⚹ centred vertically on each data point. The vertical size of these bars has been chosen to be a generous 10% of the value of the individual point. There is a negligible error in the measurement of the time and so there are no horizontal error bars.

- The line has been drawn so that it falls within each error bar. If there were vertical and horizontal error bars then the line would be drawn, if possible, to lie within the boxes formed by the bars.

- The line should be drawn so that there is an equal balance of points each side of the line as usual.

You can use error bars to give a realistic sense of the errors in your measurements when you use graphs to present your data.

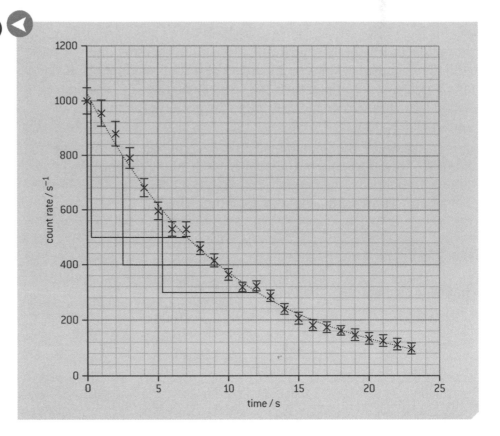

## Question

5  The activity of a sample of a radioactive nuclide was measured every three hours. The table shows the measured count rates corrected for the background.

| Time (h) | 0 | 3 | 6 | 9 | 12 | 15 |
|---|---|---|---|---|---|---|
| Correct count rate ( s⁻¹) | 1220 | 670 | 360 | 198 | 106 | 59 |

Determine the half-life of the nuclide.

## 5.4 The Standard Model

Observations of beta decay showed that there was more to the atom than protons, neutrons and electrons that never change. One way to interpret beta decay is to describe it as one neutron transforming into one proton with the emission of an electron from the nucleus. This keeps the overall charge of the nucleus constant. Similarly, positron emission can be interpreted as the conversion of a proton to a neutron with the emission of a positively charged particle.

Another such observation is the widespread existence of antiparticles. For example, the antiproton was discovered in 1955 at the University of California by Segrè and Chamberlain. By the late 1960s over 300 distinct particles had been discovered and a classification was being attempted by nuclear scientists, leading to the Standard Model.

One suggestion in the new model was the existence of a new particle called a **quark**. Experimental evidence for these was then uncovered over a lengthy period from 1977 (when the presence of the bottom quark was inferred) to 2012 (when the likely existence of the Higgs boson was confirmed).

The essential hypothesis of the Standard Model is that there is a family of quarks, given odd names: the up, down, top, bottom, strange and charm quarks. We will focus on the up and down varieties here. These quarks can be combined in either pairs or in threes. No other combination is possible and lone quarks have never been observed.

**DP ready** **Nature of science**

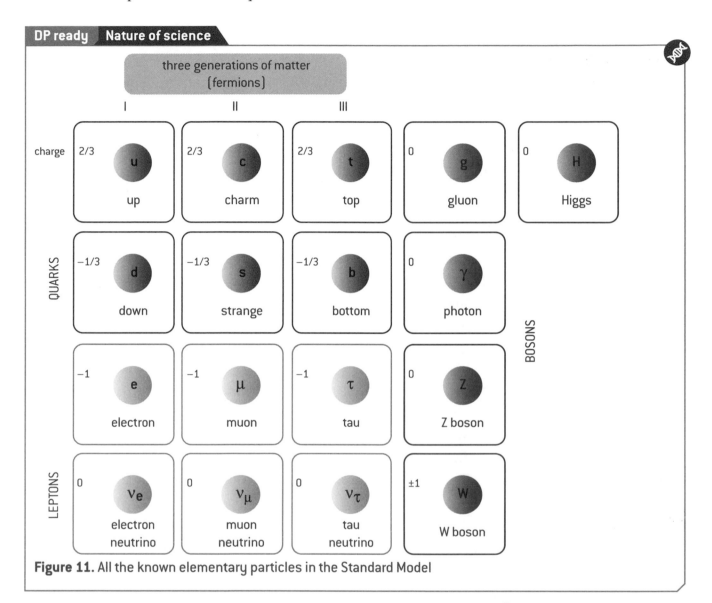

**Figure 11.** All the known elementary particles in the Standard Model

**DP ready** **Theory of knowledge**

**Science or stamp collecting?**

Is this real science? Quarks have never been directly observed and according to the Standard Model never will be, because theory suggests that as energy is added to a nucleon to tear it apart, new particles made of combinations of quarks must be formed rather than the disintegration of the original particle.

What is the significance of a scientific theory that suggests within itself that one of its main tenets is unobservable?

Beta decay can now be thought of as the change of just one of the quarks in the nucleon causing the neutron–proton conversion. Protons and neutrons are in the class of particles called baryons, which have three quarks (two-quark particles are called mesons). A proton is thought to consist of two up quarks and a down quark, the neutron

**DP link**

You will learn about the Standard Model in more detail, including leptons and bosons, in **7.3 The structure of matter** in your Diploma Programme Physics course.

If you want to know more about the Standard Model now, there are many websites available. One of the best is the CERN website (the European Organization for Nuclear Research), which has a section for students.

of two down quarks and an up quark. If a down quark in a neutron converts to an up quark, the neutron becomes a proton. The charges and other properties of the quarks also fit this suggestion. The production of the proton in beta decay is accompanied by emission of an electron and an electron antineutrino for conservation of nuclear properties—all well explained in the Standard Model.

## Chapter summary

Make sure that you have a working knowledge of the following concepts and definitions:

- ☐ Ionizing radiation is the loss of one or more atomic electrons from an atom.
- ☐ Rutherford scattering of the alpha particle led to the Rutherford model of the atom.
- ☐ The atom has a small nucleus surrounded by atomic electrons.
- ☐ The nucleus consists of protons and neutrons; these can be regarded as composed of fundamental particles known as quarks.
- ☐ Quarks are part of the Standard Model of matter.
- ☐ Atoms are defined by proton number and nucleon number.
- ☐ Isotopes are nuclides with the same proton number and different nucleon number.
- ☐ Nuclear changes that take place during radioactive decay include alpha, beta and gamma emission.
- ☐ Alpha and beta particles and gamma radiation have distinct identities and properties.
- ☐ The neutrino was predicted as one outcome of beta-minus ($\beta^-$) decay.
- ☐ Antimatter is made of antiparticles with opposite charges to their particles, and the two annihilate when they meet.
- ☐ Ionizing radiation can be detected by Geiger–Müller tubes and cloud chambers.
- ☐ Background radiation has various causes and must be taken into account during measurements.
- ☐ The becquerel is the SI unit of activity.
- ☐ There are techniques for the safe use of radioactive material in practical work.
- ☐ Half-life is constant for each radioactive nuclide and can be determined practically.

## Additional questions

1. Tritium-3 is an isotope of hydrogen. It decays by beta decay to an isotope of helium.
   a) Deduce the composition of a tritium-3 atom.
   b) State the difference between a hydrogen-1 nucleus and a tritium-3 nucleus.
   c) Write the nuclear equation for the decay of tritium-3.

2. Dosimeters are used to check the exposure to radiation of someone who works in the nuclear research industry. One older form of dosimeter is a film badge.

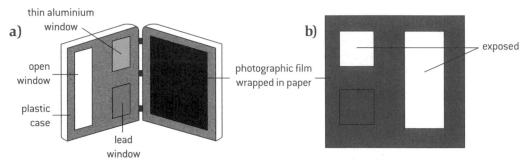

a)
thin aluminium
window
open
window
plastic
case
lead
window

b)
exposed
photographic film
wrapped in paper

**Figure 12.(a)** An opened film badge; **(b)** the developed film

Deduce the radiation to which the worker wearing the badge in figure 12 has been exposed.

3. Disposable syringes are sterilized using radiation. They are sealed in a thick plastic bag before being exposed to high levels of radiation.

   a) Suggest a type of radiation that will be suitable for sterilizing the syringes.

   b) Explain the process.

4. a) Radium-226 decays to form an isotope of radon with the emission of an alpha particle.

   i) Complete the nuclear equation.

   $$^{226}_{88}\text{Ra} \rightarrow \underline{\phantom{-}}\text{Rn} + \underline{\phantom{-}}\alpha$$

   ii) Explain what is meant by *isotopes*.

   b) Geiger and Marsden used the decay above to produce the alpha particles for their experiment.

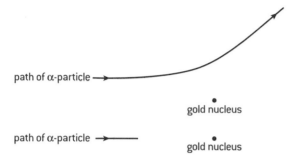

   i) The upper diagram shows the path of an alpha particle that is scattered (deflected) by a gold nucleus. Suggest the effect on the path of the particle of using a more massive isotope of gold.

   ii) The lower diagram shows an alpha particle approaching a gold nucleus along the line of their centres. Predict the path of the alpha particle after it interacts with the gold nucleus.

5. Carbon-14 is a radioactive nuclide with a half-life of 5700 years. A sample containing 1.0 mg of carbon-14 has an initial corrected total count rate of 23 counts per day.

   a) Calculate, in Bq, the activity of the carbon-14.

   b) Plot a graph showing the activity of the carbon-14 as it varies over 25,000 years.

6. The graph shows the variation in activity with time for a radioactive sample.

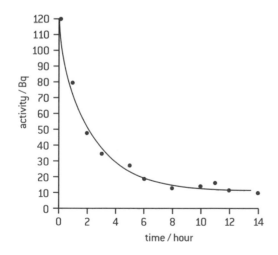

   a) i) Explain why the data points do not lie on a smooth curve.

      ii) Explain the readings obtained when the time is greater than 8.0 hours.

   b) Determine the half-life of the sample.

7. The chart shows the main sources of background radiation on Earth.

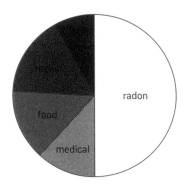

a) i) State **one** natural source of background radiation.

   ii) State the source that contributes least to background radiation.

b) A Geiger–Müller tube and counter record 1200 counts over 10 minutes with no source close to the tube. A source is then placed close to the window of the tube and the count rate is recorded. The experiment was then repeated with a sheet of card and a sheet of aluminium separately between the source and the tube. The source–tube distance remained constant.

| Absorber between source and tube | Observed average count rate ($s^{-1}$) |
| --- | --- |
| air | 33 |
| sheet of card 1 mm thick | 20 |
| sheet of aluminium 2 mm thick | 2 |

Use the results in the table to deduce all possible radiations emitted by the source.

8. Technetium-99m is an unstable atom that emits only gamma radiation as the technetium nucleus loses energy. It is used in medical imaging for investigations inside patients. The half-life of technetium-99m is 6.0 hours.

a) Explain why a source of alpha radiation would not be suitable for internal investigations.

b) i) Calculate the time it will take for the activity of a sample of technetium-99m to fall to $\frac{1}{16}$ of its initial value.

   ii) Estimate the activity of the sample after one week.

c) Outline the steps that a hospital might take to ensure that hospital staff are not exposed to radiation from the technetium unnecessarily.

9. The diagram shows part of the sequence of decays involving radon (Rn), a gas that forms naturally in rocks. Each symbol gives details of the nuclide formed and its half-life.

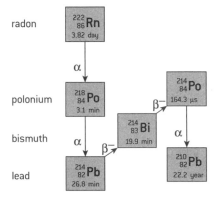

a) Explain, using examples from the diagram, what is meant by isotopes.

b) State the numbers of the nucleons and electrons in a neutral atom of bismuth.

c) Explain, using the diagram, why breathing radon can be harmful to the lungs.

10. The half-life of sodium-22 ($^{22}_{11}$Na) is 2.6 years. Sodium-22 decays to neon (Ne) by positron ($\beta^+$) emission.

    a)  Write down the nuclear decay equation for sodium-22.

    b)  i)  Explain why the positron emitted in the decay can have a range of energies.

        ii) Explain why the emitted positron is very short lived.

    c)  A sample of sodium-22 has an initial activity of $7.8 \times 10^5$ Bq. Calculate the activity of this sample after 13 years.

11. The graph shows the variation of activity of a radioactive element with time.

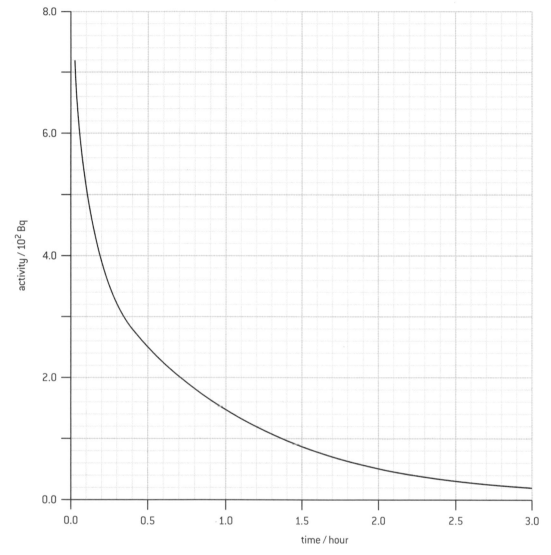

    a)  Outline what is meant by isotopes.

    b)  Deduce, using the graph, that there is more than one isotope in the sample.

    c)  The half-life of one isotope is less than 10 minutes. Estimate the half-life of the other isotope.

# Generating and using energy

> *Saving our planet, lifting people out of poverty, advancing economic growth... these are one and the same fight. We must connect the dots between climate change, water scarcity, energy shortages, global health, food security and women's empowerment. Solutions to one problem must be solutions for all.*

**Ban Ki-moon, Secretary-General to the United Nations 2007–2016, in an address to the 66th General Assembly at the UN (2011)**

## Chapter context

This chapter describes the processes involved in the **transfer** of **thermal energy**, and then discusses some of the ways that energy is generated and how this affects the Earth's **resources**.

## Learning objectives

In this chapter you will learn about:

→ the **transfer** of **thermal energy**

→ the importance of the **Sun** to life on Earth

→ the mechanisms of **nuclear fusion**

→ **solar photovoltaic cells** and solar **heating panels**

→ the principles of **thermal generating stations**

→ **nuclear fission** and the safe handling of nuclear waste

→ **fossil fuels**

→ **geothermal resources**

→ **water-, tide-** and **wind-based resources**.

## 🔑 Key terms introduced

→ conduction, convection and radiation

→ nuclear fusion

→ binding energy

→ greenhouse effect

→ Sankey diagram

→ nuclear fission

→ chain reaction and critical mass

---

### DP link

You will learn about the transfer of thermal energy when you study **8.1 Energy sources** in the Diploma Programme Physics course.

### Internal link

These three states of matter are discussed in **3.1 States of matter**, along with internal energy, heat and temperature.

## 6.1 Transferring thermal energy

**DP ready    Approaches to learning**

Many areas of physics come together in the real-life engineering applications discussed here. To achieve success in your IB Diploma Programme Physics course you will need this all-round approach to your learning.

In all three states of matter, the particles (atoms and molecules) are always moving. The degree of movement differs, however.

- The particles in a solid do not change their relative position but vibrate "on the spot".
- The particles in liquids do move around their neighbours to some degree.
- The particles that make up a gas are in free movement and are not connected to each other.

This internal motion of matter also explains the ways in which energy can be transferred within and between materials. There are three separate processes that can be identified for this transfer: *convection, conduction* and *radiation*.

## Convection

Convection is important only in fluids (liquids and gases). When part of a fluid is heated, it expands and flows upwards through the denser, cooler fluid around it. This process is convection. Because relative movement within the bulk material is not possible in solids, they cannot undergo convection.

**Practical skills: Convection experiments**

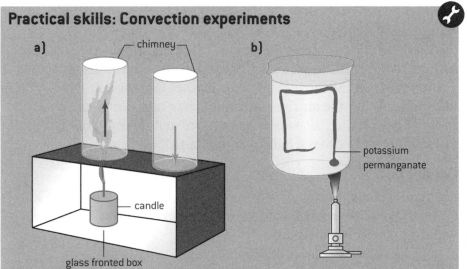

**Figure 1.** Demonstrations of convection (a) in gas; (b) in a liquid

In figure 1(a) a burning candle heats the air close to the flame. The air expands and becomes less dense than the surrounding air. It rises up the chimney. As a result, cooler air is pulled down the right-hand chimney to the flame where it also heats up, expands and rises. Smoke can be used to mark this movement.

Figure 1(b) shows the similar process of convection in a liquid. Place a small crystal of the soluble compound potassium permanganate, which is strongly coloured, on one side of the bottom of a glass beaker that contains cold water. Gently heat underneath the crystal with a small Bunsen burner flame. The heated water expands and a **convection current** is set up in the liquid, made visible by the dissolved potassium permanganate.

> **Key term**
>
> **Convection** is the transfer of energy through the movement of a fluid as a result of density changes.
>
> **Conduction** is the transfer of heat (as kinetic energy) when electrons and other particles in an object collide.
>
> **Radiation** is the transmission of energy in the form of electromagnetic waves.

> **Internal link**
>
> Density was defined in 3.1 States of matter.

The many natural examples of convection currents include convection in rocks to create new land mass on the Earth. Figure 2 shows what is happening at the bottom of the Atlantic Ocean at the mid-Atlantic ridge. The Earth's core is at a high temperature, and as a result the rock in the upper mantle is heated from below. This rock is molten and so because of convection currents wells up to reach the surface under the sea. This rock then solidifies, only to be forced outwards by more rock arriving from below. This has the effect of forcing Africa and Europe in one direction and the Americas in the opposite direction. Over geological time these convection currents have moved the east coast of South America and the west coast of Africa apart, as can be seen through the similarity in shape of their shorelines.

At the other edges of the tectonic plates that have been formed by this process, one plate is being forced under another to rejoin the upper mantle.

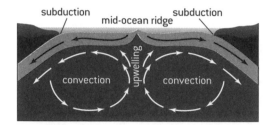

**Figure 2.** Convection currents at the mid-Atlantic ridge

Another important convection effect is the movement of wind over the Earth's surface. One of the prime reasons for this is the uneven heating of the Earth's surface by the Sun. This leads to differences in air pressure and the movement of air masses that we call wind. Where the surface is hot the air rises, creating a low-pressure area. Where the surface is cooler, air falls to the surface and the pressure is higher. These differences drive winds over the Earth's surface.

## Conduction

We should distinguish between electrical conduction (described in Chapter 2) and **thermal conduction**. For the rest of this section we will drop the word "thermal".

Conduction is important principally in solids. In liquids and gases convection plays a larger role as the links between particles are weaker. As in electrical conduction, some conductors are good, others poor. Good conductors include metals (although there is considerable variation between good and not so good); poor conductors include glass and some plastics.

### Practical skills: Conduction experiments

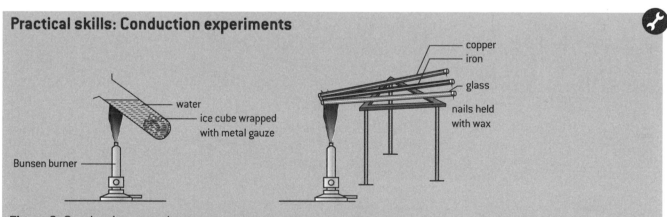

**Figure 3.** Conduction experiments

Figure 3 shows two demonstrations of conduction. Trap some pieces of ice with gauze in the bottom of a test tube of water (figure 3(a)). The water will soon cool so that the ice and water are in thermal equilibrium. Now heat the water at the **top** of the tube as shown. The water at the top will boil but the ice will not melt, showing that water is a very poor conductor. When the ice is at the top and the water is heated at the bottom of the tube, convection currents in the liquid soon help to melt the ice.

Not all solid materials conduct to the same extent. Take three rods as shown in figure 3(b) and fix nails or drawing pins to the ends of the rods with wax. The best conductor will lose its pin first.

Like electrical conduction, conduction in a metal is due to the movement of the free electrons, but there is also a contribution from atomic vibration. When a metal is at a high temperature, the fixed ions have a high average speed of vibration about their fixed position. Collisions between ions transfer this energy to regions of the metal where the ions have a lower average speed, that is, to a cooler region. The process continues until there is a thermal equilibrium where the energy transferred **from** an ion to others is equal to the energy transferred **to** an ion from those around it.

The movement of free electrons in a metal can also contribute to conduction in these materials as high-energy electrons "diffuse" their energy to cooler regions.

### Radiation

Thermal radiation is the transfer of energy via electromagnetic radiation. Above absolute zero, all atoms and molecules have some thermal motion. The acceleration of the moving charges leads to the emission of electromagnetic radiation.

**Internal link**

You can remind yourself of the role of free electrons in electrical conduction in **2.1 Electric fields and currents** and the idea of conservation of momentum in **1.4 Momentum and impulse**.

---

**Practical skills: Radiation experiments**

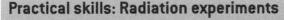

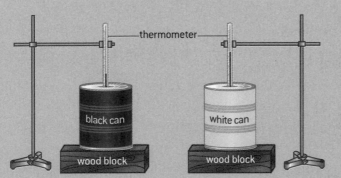

**Figure 4.** Radiation from black and white surfaces

Take two identical cans; paint one white and the other black. Fill them with identical volumes of hot water at the same temperature and add thermometers. Make sure that no radiation from one can is incident on the other can. Record the temperatures from both cans at regular time intervals and plot temperature against time. These graphs are called cooling curves. What are the differences between the cooling curves for the two cans?

Place a radiant heater between two metal plates, one painted black, the other left shiny. Thermometers attached to the rear of each plate will allow you to plot heating curves (again, temperature against time) for both plates.

**Internal link**

The properties of electromagnetic radiation were covered in **4.2 Physics of light** and **4.3 Electromagnetic radiation** and the thermal motion of particles in **3.1 States of matter**.

---

These radiation experiments suggest that black surfaces are good at both radiating and absorbing energy. White surfaces are good reflectors but poor absorbers and radiators. Hot drink dispensers are usually shiny so that they can retain the energy of the drinks for as long as possible. In hot countries, buildings are often painted white to keep them cool.

### The Dewar flask

Scottish physicist James Dewar invented the Dewar flask (or vacuum flask) as a piece of laboratory equipment in 1892. Vacuum flasks are now commonly used to keep materials (often food or drinks) hot or cold. They illustrate how, in most materials, energy transfer is due to more than one of the three heat-loss processes.

Use the internet to research the Dewar flask and find out how it is designed to reduce energy transfer.

The amount of heat lost every second through radiation depends on other factors too:

- larger surface area – greater energy loss
- higher surface temperature – greater energy loss.

### Worked example: Heat transfer example

**WE**

1. Sleeping bags are often used by people sleeping in a tent. Some sleeping bags have a fibre filling that reduces the heat transfer from the person inside.
    a) State the heat transfer processes that are involved.
    b) Explain how the bag and the fibre filling helps to reduce heat transfer.

*Solution*

a) Convection and conduction are the processes involved. (Radiation is not significant in this case.)

b) Air is trapped in the strands of the fibres. Air is a good insulator (a poor conductor). The bag and the fibres prevent large movements of the air so that convection is reduced, although it takes place to a small extent.

### Question

1 An electric kettle, made of steel, is switched on and boils a quantity of water. When the kettle is switched off it cools down.
   a) Explain how the temperature of all the water increases when the kettle is on.
   b) For the period while the kettle is cooling, describe how energy is transferred:
      i) through the kettle walls
      ii) from the exterior of the kettle.
   c) Suggest how heat transfers from the kettle surface to the surroundings can be minimized.

## 6.2 Energy resources
### Solar energy

Much of the energy used on Earth originates in the Sun. It is transmitted through the vacuum between Earth and Sun by electromagnetic radiation. The only energy resources we use that are not directly Sun-related are geothermal, nuclear and tidal (although there is an element of the Sun's gravitational field involved in tidal energy too).

This raises the question of the origin of the Sun's own energy. What is the source of the vast amounts of energy (about $4 \times 10^{26}$ J) that the Sun produces every second? (To put this value in perspective, the total energy consumption of the Earth in the whole of 2013 was estimated to be about $6 \times 10^{20}$ J.) The answer is nuclear fusion.

## Nuclear fusion

The primary fuel in a star is hydrogen, an atom with one proton and one electron. In a star the conditions are extreme. The core temperature is millions of kelvin so the electrons are stripped from the protons to form the fourth state of matter, a **plasma**. The protons move at very high speeds and so can approach very close to each other, overcoming the forces of repulsion between them. Under the right conditions, the two protons can **fuse** to form a new nuclide of hydrogen-2 (figure 5).

Hydrogen-2 must contain one proton (because it is hydrogen) and one neutron. A positron and a neutrino are also released. Through further interactions (figure 5) the new hydrogen-2 nuclei can eventually form a helium-4 nucleus (two protons and two neutrons). Overall four protons from four hydrogen-1 nuclei undergo *nuclear fusion* to produce a new helium-4 nucleus.

The remarkable thing about this change is that the total mass of the three products (figure 5) is less than the original mass of the six hydrogen-1 nuclei involved. Mass has been transferred into a different energy form. Einstein suggested in 1907 that every mass should be regarded as a store of energy, and this fusion reaction is an example of a transfer involving mass and energy.

| DP ready | Nature of science |

Einstein predicted the equivalence of energy $E$ and mass $m$ using the equation $\Delta E = \Delta m c^2$ where $c$ is the speed of light in a vacuum.

Another interpretation of the "loss" of mass is that the constituents of the helium nucleus are more tightly bound together than the original hydrogen (obviously the single proton is not bound to anything). This means that energy must be transferred into the helium to break it up. Conversely, when the helium forms from its individual protons (with two of them converting to neutrons) this *binding energy* must be removed otherwise they will be too energetic to stick together. The result is that about $10^{-29}$ kg of matter disappears when one helium nucleus forms. This mass is equivalent to $10^{-12}$ J of energy. About $10^{39}$ fusions occur each second in the Sun, producing the colossal $4 \times 10^{26}$ J quoted earlier. However, enough hydrogen remains to power the Sun for another 6 billion years!

## The greenhouse effect

The energy generated in the heart of the Sun makes its way to the surface that we see. Then, in the form of individual photons, it travels outwards at the speed of light, taking about 9 minutes to reach the Earth.

When the radiation reaches the Earth, some is trapped in the atmospheric system by the *greenhouse effect*. This effect is vital to us because without it the Earth would be about 25 K cooler. However, the levels of carbon dioxide and other greenhouse gases in the atmosphere are currently increasing, leading to an **enhanced greenhouse effect** and inevitable increases in the global average temperature. This will affect climate globally and have a serious detrimental effect on life if it is not reduced.

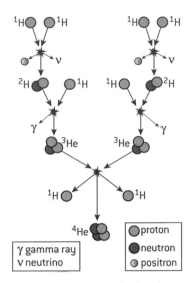

**Figure 5.** Nuclear fusion in a star

 **Key term**

**Nuclear fusion** occurs when one or more nuclei combine to form a larger nucleus with the release of energy.

**Binding energy** is the minimum energy required to separate a nucleus into its separate nucleons.

 **Internal link**

We saw in **5.2 Radioactive decay** that in beta decay a neutron changes into a proton with the emission of an electron and a neutrino and that the reverse process is possible, with a proton becoming a neutron (**5.4 The Standard Model**).

 **Key term**

The **greenhouse effect** occurs when energy from the Sun is trapped in the atmosphere of a planet. When the atmosphere is more transparent to visible light from the Sun than to infra-red radiation from the surface of the planet, the infra-red is not radiated away as fast as energy is transferred in.

### Using the Sun's energy directly

Radiation from the Sun can be used directly on both national and domestic scale by devices that transfer the Sun's energy either into an electrical form (photovoltaic cells) or into thermal energy (solar heating panels).

#### Photovoltaic cells

**Figure 6.** A solar photovoltaic cell array

A **solar photovoltaic (PV) cell** converts electromagnetic energy from the Sun into electrical energy (figure 6). The cell has a layered construction of semiconductors (with resistance midway between that of a conductor and an insulator). Both layers have free electrons, and their chemical composition is such that when photons are incident on the cell, the electrons are released and are driven through an external circuit. In this way, the PV device behaves like an electric cell, but the energy transfer is from light to electrical, not chemical to electrical.

#### Solar heating panels

The principle behind **solar heating panels** (figure 7) is quite different from that of PV cells.

The solar heating panel absorbs energy from the Sun and heats a glycol–water mixture that is circulated by a pump. (The glycol – a liquid with a low freezing point – prevents the system freezing.) The hot mixture flows in a circuit through a hot-water storage cylinder where the energy is exchanged with water alone. The hot water in the cylinder can be used to heat other water or can be used directly in a central heating system.

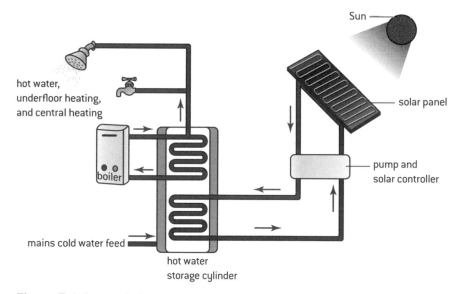

**Figure 7.** A domestic hot-water system with solar heating panels

### Worked example: Calculations involving solar panels

2. A small settlement consists of 15 houses. Solar heating panels are proposed to provide hot water for the houses. At the location of the settlement, the average intensity of the Sun at the Earth's surface is 350 W m$^{-2}$. The solar heating panels have an efficiency of 22%.

   a) Each house requires on average 4.5 kW of power for water heating. Calculate the minimum area of solar heating panel required.

   b) State the disadvantages of using solar power for this purpose.

*Solution*

a) The total power requirement is $15 \times 4.5 \times 10^3 = 6.75 \times 10^4$ W.

Allowing for 22% efficiency, the energy required each second from the Sun is $\frac{6.75 \times 10^4}{0.22} = 0.307\,\text{MW}$.

So the area required is $\frac{3.07 \times 10^5}{350} = 880$ m$^2$ (this is an area of panel about 30 m × 30 m)

b) Disadvantages include:

- no energy transfer at night
- power production fluctuates with weather conditions and cloud
- power production fluctuates with seasons
- a large area of panel required.

## Thermal power stations

Modern societies depend on a constant supply of electrical energy that has to be generated. Properly we should talk about energy transfer into the electrical form, but the term "energy generation" is so common that it is difficult to avoid. Thermally powered generating stations are used worldwide. Figure 8 shows the principles behind a thermal power station. Thermal energy is used to create steam at high pressure and at a temperature well above 373 K (it is superheated and therefore stores large amounts of energy). This steam drives a turbine connected to an electrical dynamo. As it does so, its pressure and temperature reduce. Normally, the steam will be condensed and returned to the boiler. Various sources of heat are used, principally fossil (coal, gas, oil) and nuclear fuels (nuclear fission). We will look at the processes used in the generating station itself and then examine issues connected with the fuels.

 **Internal link**

You can find a reminder of the physics of alternating current generation and the transmission of the energy in **2.4 Electromagnetic induction.**

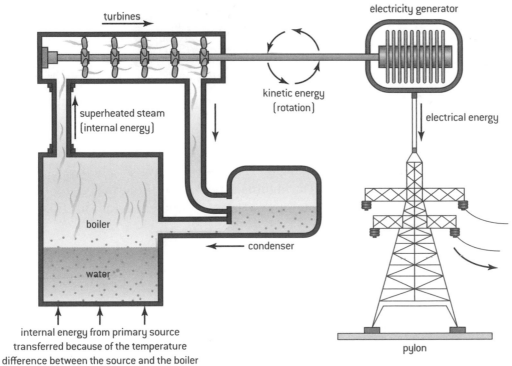

**Figure 8.** Energy conversion in a thermal generating station

## Key term

A **Sankey diagram** shows the flow of energy in a device or system. These are the rules for drawing a Sankey diagram:

- Each energy source and energy sink in the system is represented by an arrow.
- The width of each arrow is proportional to the size of the loss.
- Energy flows from left to right.
- Losses move to the top or bottom.
- The diagram can represent power as well as energy changes.

The energy transfers are depicted in figure 8: the rotational kinetic energy in the dynamo is converted into electrical energy and then transmitted, as alternating current through a grid system, to places where electrical energy is needed.

Power stations are complex pieces of engineering. There are many points at which energy loss can occur between the initial transfer of chemical or nuclear energy and its eventual use in an electrical form by the end-user. The size of these energy losses can be visualized with a *Sankey diagram*.

Figure 9 shows the Sankey diagram for a car (automobile) engine. The numbers on the diagram are percentages. The width of the arrow for input energy (from the chemical and oxygen fuel) corresponds to 100%. The various losses in the engine are given above and below the diagram: the losses in the gear train, the sound output, and so on. Finally, 23% of the original energy is available to the car. You might research the losses to this remaining energy: air resistance, internal friction in the tyres and so on. You could then extend the diagram to establish how much of the original energy ends up as kinetic energy of the vehicle.

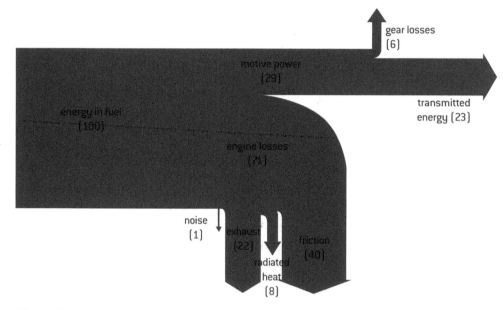

**Figure 9.** A Sankey diagram showing the energy losses in a car engine

### Question

2   A kettle is rated at 2.0 kW. It takes 1.5 minutes to boil a quantity of water. During this time 3.6 kJ of thermal energy is wasted to the surroundings. Draw a Sankey diagram for this process.

### Fossil fuels

Fossil fuels used in power stations include coal, natural gas and oil.

- Burning the fuel is straightforward, particularly gas and oil (coal needs to be crushed to powder for easy injection into a furnace).
- Significant amounts of power can be produced (the largest station in the world has an output of 6 GW).
- The fuel is cheap.

However, fossil fuels have several disadvantages too.

- They are **nonrenewable** – they take geological times to accumulate and cannot be replaced within a reasonable time.

- Burning the fuels releases carbon dioxide into the atmosphere. This $CO_2$ was locked into the fuels when the plants from which they were made were compressed in the ground. Release of this gas by power stations has had a major impact on the enhanced greenhouse effect.

- Fossil fuels have other important uses as the raw materials for plastics, medicines and other products. Burning them for energy is a waste of a limited resource.

- Fuel transportation costs are significant in some locations. This reduces the efficiency of the whole process.

Figure 10 shows a typical Sankey diagram for a fossil-fuel power station. Comparing the output and the input shows that the efficiency of a typical station is 25–40%.

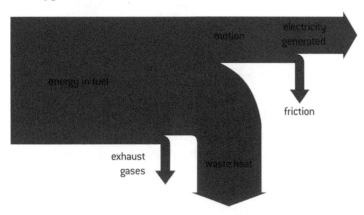

**Figure 10.** Sankey diagram for a coal-fired generating station

## Nuclear fission

Another source of heat energy for thermal power stations is *induced nuclear fission*. Earlier in this chapter we looked at the physics of fusion. Nuclear fission is a separate process that also converts mass into an energy form.

Uranium-235 is an isotope of uranium that can absorb a neutron to become a uranium-236 nuclide (figure 11). The new atom is highly unstable and soon splits into two or more nuclear fragments with the release of a few neutrons, usually two or more. These fragments are more stable, in other words, lower in energy than the original nucleus, and the surplus energy is released. It is not possible to predict the nature of the nuclear fragments (the diagram shows a barium nuclide and a krypton nuclide being produced, but many other combinations are possible).

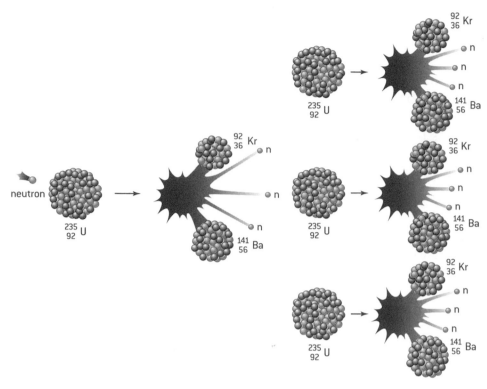

**Figure 11.** A chain reaction in uranium-235

The neutrons released in the fission can go on to interact with other U-235 nuclei and initiate what is known as a *chain reaction*. The condition for this to happen is that at least one new fission is produced by each original fission; in other words, each nuclear fission must produce one available neutron, allowing for any neutron losses in the system. There must be a certain amount of U-235 for this to happen (so that neutrons trigger fission before leaving the system); this is known as the *critical mass*. This is a self-sustaining reaction that produces a large amount of kinetic energy (due to the motion of the new nuclei and the emerging neutrons). This kinetic energy, after conversion, boils the water to produce steam for the turbines in the nuclear power station.

However simple the basic physics, the design of a nuclear reactor is not straightforward. There are some aspects to the process that require care on the part of the designers.

- The end products, including the fission fragments, are highly radioactive and need to be kept away from living organisms once removed from the reactor.

- The neutrons that emerge from the fission event are travelling at very high speeds. Slower neutron speeds are best for initiating the fission. The neutrons need to be slowed down.

- The reactor must have a means of control, both to vary the output and to maintain its safety.

Figure 12 shows a schematic design of one type of nuclear reactor. It shows the features that are required to enable the reactor to function efficiently and safely: the **moderator** and the **control mechanism**.

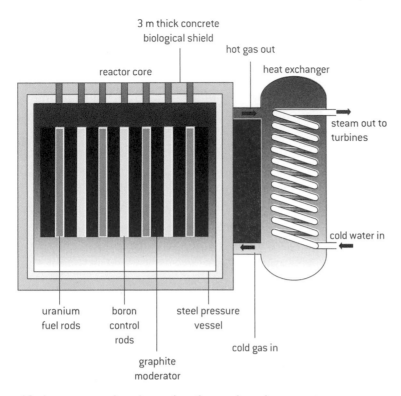

**Figure 12.** A cross-section through a thermal nuclear reactor

## Moderator

A simple piece of mechanics is involved in slowing down the neutrons. They enter the **moderator**, which must be a material that does not absorb neutrons. The fast-moving neutrons collide with atoms in the moderator. These collisions are elastic, so the neutrons gradually transfer kinetic energy to the thermal energy of the moderator atoms. The neutrons slow down and the temperature of the moderator increases as a result. After a relatively small number of collisions the neutron will have a kinetic energy that enables it to interact more effectively with uranium-235. The moderator is shaped so that after the appropriate number of collisions, neutrons are likely to leave the moderator and return to the uranium fuel rods.

An ideal moderator would have a mass equal to the neutron as this would transfer the maximum possible amount of energy to the moderator. However, hydrogen absorbs neutrons (to form hydrogen-2) so is not suitable. Commonly used moderator materials include ordinary water, heavy water (deuterium oxide) and graphite (carbon).

## Control rods

Control rods absorb the neutrons completely in order to slow or stop the fission reaction. Nuclear engineers use elements that are good absorbers of neutrons, such as indium, boron, silver and cadmium.

## Heat exchanger

It is usual to transfer energy from inside the reactor core to the turbines through a heat exchanger. Hot liquid or gas from inside the reactor heats water in the exchanger so that any contaminated liquid or gas cannot easily be transferred to the outside.

 **DP link**

You may study the mechanics of the moderator process in **8.1 Energy sources**.

*Handling radioactive products*

As the uranium nuclei in a fuel rod undergo fission, the rod itself changes. One obvious change is that each uranium nucleus has become two new nuclei. Although they have fewer nucleons inside them, the two new nuclei take up more space in total than the original uranium. This deforms the rod, and if too much deformation occurs then the rod could become stuck in its channel. So only a small fraction of the uranium is allowed to fission before the rod is removed from the reactor. The unused uranium can be chemically extracted from the rod and placed into a new rod. The fission products are removed; they are unstable when formed and extremely radioactive. The fuel rods themselves need to be very carefully handled, so are usually handled remotely using robots.

---

**DP ready**  **Nature of science**

The statement that the two new nuclei take up more space in total than the original uranium seems odd. Why is the new volume not equal to the old given that the number of nucleons is roughly the same?

To answer this, let's assume that the density $\rho_n$ of the nuclear material (the protons and neutrons) is constant and that nuclei are spherical in shape. This means that

$$\rho_n = \frac{\text{mass of nucleus}}{\text{volume of nucleus}} = \frac{mA}{\frac{4}{3}\pi r^3} = \frac{3mA}{4\pi r^3},$$ where $m$ is the mass of a nucleon, $r$ is the radius of the

nucleus and $A$ is the nucleon number. Rearranging gives $r = A^{\frac{1}{3}} \times \sqrt[3]{\frac{3m}{4\pi}}$. In other words $r \propto A^{\frac{1}{3}}$.

Doubling the number of nucleons in a nucleus only changes the nuclear radius by a factor of $\sqrt[3]{2}$, which is about $\frac{5}{4}$. So splitting one nucleus into two and losing only two or three neutrons in the process means that the space occupied by the two nuclei is larger than the original.

---

After extraction from the reactor core, the fuel rods are usually placed in a storage tank of liquid while the most active (in other words, short-lived) nuclei decay before any further manipulation of the spent rod is carried out.

Nuclear reactors also produce large quantities of low-activity waste that poses little risk to the workers in the nuclear plant, but strict precautions are used to prevent this material entering the environment.

Finally, other products of the reaction process include gamma rays and other particles. Exposure to these emissions is reduced using thick layers of concrete around the pressure vessel that encloses the active materials.

---

## Worked example: Calculations involving nuclear fission

3.  A possible fission reaction is $^{235}_{92}\text{U} + ^{1}_{0}\text{n} \rightarrow ^{141}_{56}\text{Ba} + ^{92}_{36}\text{Kr} + x^{1}_{0}\text{n}$.
    Each fission of uranium-235 produces $2.8 \times 10^{-11}$ J of energy.

    **a)**  State the value of $x$.

    **b)**  A nuclear generating station that uses $^{235}_{92}\text{U}$ as its fuel has a power output of 18 MW at an efficiency of 35%.

    Determine the mass of fuel used every day in the station.

*Solution*

a) Mass numbers: $235 + 1 = 141 + 92 + x$

$x = 3$

b) The total power required, taking efficiency into account, is $\dfrac{18 \times 10^6}{0.35} = 51.4$ MW. The number of fissions required every second is $\dfrac{51.4 \times 10^6}{2.8 \times 10^{-11}} = 1.84 \times 10^{18}$.

Converting to moles, this is $\dfrac{1.84 \times 10^{18}}{6.02 \times 10^{23}} = 3.05 \times 10^{-6}$ mol s$^{-1}$.

1 mol U-235 has a mass of 0.235 kg, so mass required is

$\quad 3.05 \times 10^{-6} \times 0.235 = 7.17 \times 10^{-7}$ kg s$^{-1}$.

Converting to one day yields

$\quad 7.17 \times 10^{-7} \times 24 \times 60 \times 60 = 0.062$ kg day$^{-1}$.

## Geothermal resources

In some parts of the world there are active volcanoes or rifts in continental plates. These offer alternatives to the use of fossil fuels or nuclear material. At such places, very hot water can be found close to the surface, the water being heated as it passes through the hot rocks. "Close" is a comparative term. In Iceland, one of the principal users of geothermal energy, a borehole 4.7 km deep is being evaluated for the transfer of energy. The bottom of the borehole has temperatures of at least 430°C, and one hole is thought to be capable of generating around 40 MW of power. Iceland straddles the mid-Atlantic ridge and evidence for the instability of the geology along this ridge is seen everywhere in the country.

To extract energy effectively, water is used in a cycle (figure 13). First, cool water is injected into the ground at a point above the hot rocks. Underground, the water boils to form steam and, under pressure, is fed to a turbine at the surface for electricity generation. The water cools as it transfers its energy and can be returned to the cycle.

 **Internal link**

The geothermal processes at the mid-Atlantic ridge are described in 6.1 Transferring thermal energy.

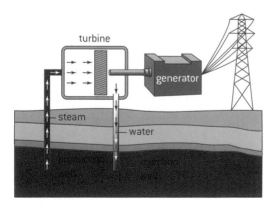

**Figure 13.** Geothermal energy extraction

Geothermal energy is also used for heating homes and producing hot water directly.

## Renewable resources

So far we have referred to turbines as propeller-like devices rotated on an axis by steam pressure. There are other ways to power a turbine, too, either wind-based or water-based (including wave and tidal).

### Wind power

Wind turbines have a rotor blade shaped like a propeller that transfers the kinetic energy of wind directly into an electrical form. We will concentrate on the type known as horizontal-axis turbine (figure 14(a)) here, but there are other types in which the turbine rotates about a vertical axis (figure 14(b)).

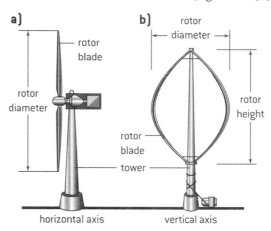

**Figure 14.** Wind turbines with (a) horizontal axis, (b) vertical axis

In the horizontal-axis turbine, wind blows in (from the left of figure 14) and transfers its kinetic energy to rotational kinetic energy of the rotor blade. The rotor turns the generator (dynamo) through a gearbox so that charge flows from the generator to an external circuit.

To calculate the maximum amount of momentum that can be transferred from wind to turbine, call the wind speed $v$ and the length of each rotor blade $r$. Now imagine a cylinder of air that represents the mass of air that moves into the turbine every second (figure 15). This cylinder has a radius $r$ and a length $v$. Its volume is the area of the cylinder face × cylinder length, which is $\pi r^2 \times v$. So the mass of the air in this cylinder is $\rho \times \pi r^2 \times v$ where $\rho$ is the density of the air. We will assume for now that all the kinetic energy of the wind is transferred to the turbine (returning to this point later). The kinetic energy arriving at the turbine is $E_k = \frac{1}{2} mv^2 = \frac{1}{2} \rho \pi r^2 v \times v^2$ or $E_k = \frac{1}{2} \rho \pi r^2 v^3$.

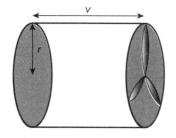

**Figure 15.** The cylinder of air that drives a wind turbine in 1 second

This is the theoretical maximum energy that the wind can deliver. However, it is not the maximum energy that the turbine can gain. We have assumed that the air stops moving completely once it has passed through the turbine, giving up all its kinetic energy. This cannot happen in reality, as there would be an accumulation of air molecules at the turbine. In fact, the air has to keep moving in order to clear the turbine, so the change in wind speed is somewhat less than $v$.

Albert Betz, a German scientist, showed about a century ago that the maximum kinetic energy that can be derived from a wind turbine is $\frac{16}{27}$ of the kinetic energy of the wind. Most practical turbines reach 70–80% of the Betz limit.

## Worked example: Calculations involving wind power

**4.** A wind generator produces 12 kW of power when the speed of the wind is 7.0 m s$^{-1}$.

    **a)** Calculate the power output of the generator when the speed of the wind is 14 m s$^{-1}$.

    **b)** State any assumptions that you made in the calculation.

### Solution

**a)** The energy output is directly proportional to (wind speed)$^3$. So doubling the speed will increase the energy transfer by $2^3 = 8$. The energy output is 96 kW.

**b)** The assumption is that the efficiency of the turbine remains constant (unlikely).

## Question

**3** The graph shows the variation of the maximum energy available from the wind with wind speed.

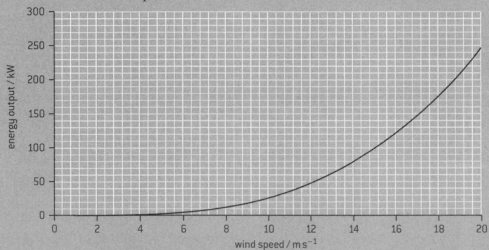

The turbine can deliver 70 % of the energy removed from the wind.

Estimate the power output of this wind turbine at a wind speed of:

    **a)** 9.0 m s$^{-1}$    **b)** 18.0 m s$^{-1}$.

### Water power

Another common way to use a turbine is to make water do the turning. In this case, water moves through a turbine at speed $v$, and some fraction of the initial kinetic energy of this water can be made available to the turbine, which as usual is connected to a dynamo.

The wind-turbine equation given on page 156 can also be used to calculate the maximum kinetic energy available from the water. So, this time we focus on the processes that cause the water to move rather than the mechanics of the conversion itself.

Two of the water-power systems involve water that flows from a high level to a low level, thus transferring from gravitational potential energy to (finally) an electrical form. These are **hydroelectric systems**.

Water is stored (figure 16), generally behind a dam across a river, and released through a channel down towards a turbine and dynamo. After passing through the power station the water continues as part of the river.

The water drops through a vertical distance of $h$. Each kilogram of water has an initial gravitational potential energy of $1 \times g \times h$; assuming no energy losses this will be converted to $\frac{1}{2} \times 1 \times v^2$, where $v$ is the speed of the water and is $\sqrt{2gh}$.

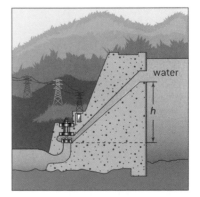

**Figure 16.** A hydroelectric system

When $m$ kilograms of water pass through the turbine each second, the maximum power available from the water will be
energy of 1 kg of water × mass of water through turbine per second

$$= \frac{1}{2} m \times (2gh).$$

The drawback to a hydroelectric system of this type is that it depends on the availability of water in the reservoir. In times of drought this can be problematic. It is also usually necessary for some water flow to be maintained at all times so that the river has water for people living downstream and also for fish in the river.

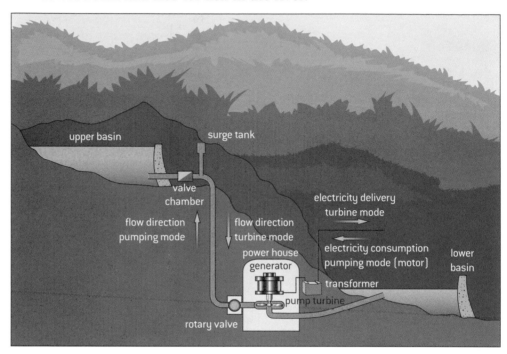

**Figure 17.** A pumped storage system

Pumped storage systems are a variant of the hydro-electric system but have more flexibility. There are two reservoirs (figure 17), one upper and one lower. The upper reservoir is filled naturally by rivers and rain water. Water runs from the upper reservoir to the lower through a turbine and generator, just as before, at times when demand for energy is high. However, unlike the hydroelectric scheme (Figure 16), when demand is low and energy is cheap, the water can be pumped back up to the higher reservoir by reversing the action of the dynamo so that it becomes a pump.

Pumped storage systems can be turned on very quickly, usually within seconds, and are used to generate energy when the demand peaks.

### Wave and tidal power

Two more water-powered generating systems rely on the ocean tides for their operation. In some parts of the world, tides produce significant changes in water level twice every day. (The tides are driven by the combined effects of the gravity from the Moon and Sun acting on the large bodies of water on Earth.) These water level changes can be used to generate electrical energy.

As the tide moves in and out, submerged turbines are rotated by the water flow. In a barrage system (figure 18), large barriers can be raised

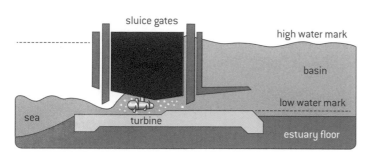

**Figure 18.** A tidal barrage system

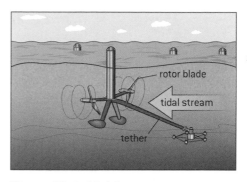

**Figure 19.** A tidal flow system

or lowered to control the flow. Tidal flow generators (figure 19) are simply placed in the region where the tide flows. There is no attempt to control the rate.

As with the wind turbine, the equation $E_k = \dfrac{1}{2}\rho\pi r^2 v^3$ gives the kinetic energy available from the water. It is interesting to compare wind and tidal stream generators in terms of their output. The density of water is roughly 800 times greater than that of the air. For even relatively small water speeds and modest turbine-blade radii, substantial amounts of electrical energy can be generated using a water-based system.

There are several ways in which the kinetic and gravitational potential energy of surface waves on the sea can be harnessed for electrical energy conversion. One technique (figure 20) forces the incoming waves to enter a chamber. As they do so, the air above the waves is compressed and forced through an air turbine. This turns the turbine and therefore a generator in the usual way. The system is designed so that the turbine rotates in the same direction independent of the air flow direction (this is called a Wells turbine) and so energy is transferred whether the water is compressing or expanding the air in the chamber.

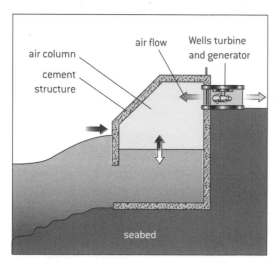

**Figure 20.** The Wells turbine for harnessing wave energy

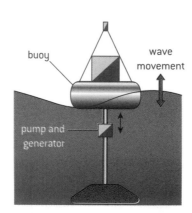

**Figure 21.** Floating buoy for harnessing wave energy

A second technique is to anchor a generator to the sea bed (figure 21). The generator is fixed to a buoy floating on the water surface. As the water surface moves up and down the generator is turned by a gear mechanism and transfers the kinetic energy of the buoy into electrical energy.

Once again, we can estimate the maximum power available from a wave. We will model the wave (figure 22) as having a square cross-section of overall height $2A$ and wavelength $\lambda$. The wave speed is $v$ and the wavefront of the wave is $L$. The density of the water is $\rho$.

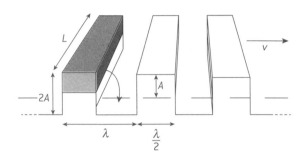

**Figure 22.** Modelling a wave as a square wave

As the crest of the wave (shaded in one wave on the diagram) falls into the trough, it loses its gravitational potential energy and the water surface will become flat. The distance fallen by the centre of mass of the upper half of the wave is $A$. So this change in gravitational potential energy is $mgh$ as usual, which is $(A \times \dfrac{\lambda}{2} \times L \times \rho) \times g \times A$.

This is the energy that can be obtained from **one** crest, so we now need to know how many waves arrive in one second. From wave theory, it is the frequency of the wave: $f = \dfrac{v}{\lambda}$.

The maximum energy available from all the crests that arrive in one second is $mgh \times f$, which is $(A \times \dfrac{\lambda}{2} \times L \times \rho) \times g \times A \times \dfrac{v}{\lambda}$. This simplifies to $\dfrac{1}{2}\rho g A^2 L v$, and because this is the energy arriving per second it is also the maximum power available from the whole wave. However, it is extremely unlikely that a wave generator can sample the whole wave and it is more useful to quote the maximum power available from a one-metre length of the wave. This is $\dfrac{1}{2}\rho g A^2 v$. The unit for this quantity is W m⁻¹.

**Internal link**

You can find the relevant mechanics in **1.3 Work and energy** and the wave theory in **4.1 Waves in theory** to support this proof. The definition of density is in **3.1 States of matter.**

---

**DP ready**   **Nature of science**

**Pulling things together**

Notice how many separate areas of physics are brought together in this relatively simple calculation. Always be alert for the connections within the subject and be prepared to transfer your knowledge from one area to another.

---

**Worked example: Calculations involving wave power**

5.  A wave of amplitude 0.80 m is moving towards the shore at 2.7 m s⁻¹. The wave is incident on a wave-energy converter of input width 5.7 m. Calculate the maximum power that can be converted by the generator. (Density of seawater = 1030 kg m⁻³)

*Solution*

The maximum power for every linear metre of wave is

$\dfrac{1}{2}\rho g A^2 v = 0.5 \times 1030 \times 9.81 \times 0.80^2 \times 2.7 = 8.7$ kW.

The total maximum power available is 50 kW.

## Chapter summary

Make sure that you have a working knowledge of the following concepts and definitions:

☐ There are three mechanisms for thermal energy transfer: conduction, convection and radiation.

☐ Nuclear fusion in the Sun combines hydrogen nuclei into helium nuclei.

☐ Solar energy is directly converted using photovoltaic cells and heating panels into energy for use.

☐ Electricity is generated in thermal generating stations using steam turbines:

  ○ by burning fossil fuels

  ○ by nuclear fission.

☐ Sankey diagrams are used to visualize energy losses.

☐ Geothermal energy is generated using steam heated by deep rocks.

☐ Wind turbines work by converting a part of the wind's kinetic energy.

☐ Water-based techniques for energy generation include:

  ○ hydroelectric systems

  ○ pumped storage systems

  ○ tidal systems

  ○ wave-based systems.

## Additional questions

1. Explain the advantages of:

   a) a copper cooking pan with a plastic handle

   b) wooden doors over metal doors

   c) using a dry cloth to pick up a hot plate rather than a wet cloth.

2. Outline why hot-water storage tanks are often lagged with a fleece cover. In your answer you should consider all three modes of thermal energy transfer.

3. Explain why a solar heating panel normally has a black layer behind the heating pipes.

4. Uranium-235 can spontaneously fission to form nuclides of barium-141 and krypton-92 without the presence of an initiating neutron.

   a) Complete the nuclear equation for this fission:
   $$^{235}_{92}U \rightarrow \,^{\phantom{92}}_{36}Kr + \,^{141}_{\phantom{-}}Ba + x\,^1_0n$$

   b) The mass of the products after the fission is less than the original mass by $3.1 \times 10^{-28}$ kg.

      i) Show that this is equivalent to an energy of about $3 \times 10^{-11}$ J.

      ii) Calculate the number of U-235 nuclei that must spontaneously fission to transfer 1.0 kJ of energy.

   c) Outline how this fission reaction can lead to a chain reaction.

5. A 500 MW generating station uses natural gas for fuel; when one kilogram of this gas is burnt, an energy of 56 MJ is transferred into thermal form. The efficiency of the power station is 29%.

   Calculate the rate at which natural gas is used in the generating station.

6 a) Outline what is meant by the greenhouse effect.

   b) i) Describe what is meant by the enhanced greenhouse effect.

      ii) Explain why the enhanced greenhouse effect may increase the temperature of the Earth's surface.

7. A coal-fired generating station has an output of 4.0 GW. Determine the minimum mass of uranium-235 that is required by a nuclear generating station to provide this power output for one year (1.0 kg of uranium-235 can release an energy of 82 TJ).

8. Use the following data to determine the change in mean sea level as a result of the entire Antarctic ice sheet melting into the Earth's oceans.

Density of water = 1000 kg m$^{-3}$
Density of ice = 920 kg m$^{-3}$
Area of Antarctic ice sheet = $1.5 \times 10^7$ km$^2$
Mean thickness of Antarctic ice = 1.5 km
Area of Earth's oceans = $3.5 \times 10^8$ km$^2$

9. The albedo of the Earth's surface is defined as $\dfrac{\text{intensity reflected by surface}}{\text{intensity received by surface}}$. The average albedo of the Earth is 0.3. The intensity of solar radiation at the orbit of the Earth is 1400 W m$^{-2}$. Determine the average reflected intensity from the surface of the Earth.

10. A fossil-fuel generating station has an input power of 280 MW and an electrical output of 100 MW.

    a) i) Outline what is meant by a fossil fuel.

    ii) Give **one** example of a fossil fuel.

    iii) Explain why fossil fuels are nonrenewable.

    b) i) Draw a Sankey diagram for the generating station.

    ii) Use the Sankey diagram to estimate the overall efficiency of the generating station.

    c) Explain why the use of fossil fuels is widespread even though it is known that they contribute to atmospheric pollution.

11. A heater burns a liquid fuel to heat a room.

| density of liquid | 820 kg m$^{-3}$ | temperature of air entering heater | 12°C |
|---|---|---|---|
| rate at which liquid is burnt | $1.3 \times 10^{-4}$ kg s$^{-1}$ | temperature of air leaving heater | 32°C |
| energy produced by 1.0 m$^3$ of fuel | 27 GJ | specific heat capacity of air | 990 J kg$^{-1}$ K$^{-1}$ |

    a) Calculate the power output of the heater. Assume that the specific latent heat of vaporization of the liquid fuel is negligible.

    b) Determine the mass of air moving through the heater in one minute.

12. A pumped storage generating station has a height difference of 270 m between upper reservoir and turbine. The volume of the upper reservoir is $2.5 \times 10^6$ m$^3$ and the maximum flow rate for the water to the turbine is 350 m$^3$ s$^{-1}$. The density of the water is 1000 kg m$^{-3}$.

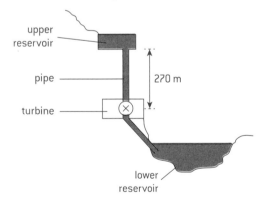

    a) i) Determine the maximum energy that can be delivered to the turbine.

    ii) Calculate the maximum power that can be delivered by the water as it flows through the turbine.

    b) Friction of water in the pipe accounts for 27% of the energy loss in the system, friction in the turbine is 18%, and electrical resistance losses are 6%.

    i) Calculate the overall efficency of the transfer from gravitational potential energy to electrical energy.

    ii) Draw a Sankey diagram of this process.

Much of the advice here is not specific to physics and can be applied to all your learning and study. Success in the Diploma Programme relies on effective and consistent study. Your teachers will expect you to be able to maintain balance between all your subjects and will not be impressed if you spend time on one subject at the expense of another.

## 7.1 Approaches to your learning

You should concentrate on developing a good all-round understanding of your physics. While knowledge and rote learning are important, a good understanding of the concepts is crucial. All areas of physics share common ideas such as conservation, energy, momentum and rate of change. Make a habit of trying to link the subject you are studying now with what you learnt earlier. The *Internal link* icon (∞) in this book shows areas where learning can be joined up.

Our brains are good at forming links – make the most of this skill. The Diploma Programme is all about learning and developing concepts across all your subjects. Aim to become an effective student physicist with good learning skills and an independent approach to your work. In the Programme you should seek to develop the following:

- *Communication skills*
  All scientists need to communicate their findings and results effectively. These important skills will pervade your entire Diploma work. As a Diploma student you will also transfer these skills to your other group subjects, the Theory of Knowledge and the Extended Essay.

- *Self-management skills*
  You need to know the right conditions for learning for yourself, and to be able to create them. There is advice on this later in this chapter.

- *Research skills*
  These will help you not just in your IA project but in collecting ideas and material for the Theory of Knowledge essay and the Extended Essay.

- *Thinking skills*
  You should develop the ability to think critically, to innovate and be creative within a scientific context.

- *Social skills*
  These are your skills at being an effective member of a group, whether in the group 4 project, in a collaboration in group 6, in a fieldwork survey in group 3 or just organizing an activity with friends.

These skills make up the five areas addressed in the IB by the phrase *Approaches to learning* (☺). Throughout your course (and beyond) ask yourself the following questions:

- What are my present skills in physics; how well are they developed?

- Are any of my skills too weak for comfort; how can I improve them?

- What additional new skills do I need; how can I learn them?

## 7.2 Good study habits

The start of a new course—whether moving to the IB Diploma from a GCSE, national qualification or Middle Years Programme background—is an appropriate time to review your study habits. A good approach to learning is vital if you are going to make the most of the programme. In the spirit of the five *Approaches to learning* skills above, review how you study and ask yourself what improvements you can make.

## Help your brain

The key to good study is to integrate all the information you are learning and to retain it. You need long-term, not short-term, memory, and the way to transfer from short to long-term is by reviewing material regularly.

Our memory of a lesson (or any other experience) is at its best immediately afterwards; then the memory starts to fade. To retain the memory longer you need to review the material about one day later. But that memory will not be permanent either. The next time for a review is after a week, and then once again after a few weeks. Continue this review process and eventually the memory will become unshakeable. Think about lessons when you were younger: the teacher set homework on the day, gave a test on the material perhaps a week later, and then had a major test on material at the end of a term or semester. This was designed to make you review the work over and over again.

Some of the advice on note-taking later in this chapter relates to this idea of review.

## Planning for study

You need to plan your work schedule if you are to review regularly as suggested above. The Diploma Programme is demanding of time. You take six subjects and Theory of Knowledge classes together with other school commitments and your creativity, activity and service (CAS) projects. Then there is your social and family life outside school. You need to use a planning diary into which you put **all** your commitments. Later, it will be important to include the revision you intend to undertake. There is advice about this later.

## When and where to study

Everyone is different. Some people work best in the morning; others peak in the evening. Play to your strengths here; don't force a work pattern on yourself that your brain and body say is wrong. If you are a morning person, get up early and do your work before school.

Include breaks in your planned schedule – remember how the brain operates. Everyone needs 10–15 minutes to settle down to effective study, so don't plan to do any important work then. Equally, you need breaks of a few minutes at regular intervals, perhaps every 20–40 minutes. But do not take too long over the break as then you will need settling time again. Making and drinking a cup of coffee is about the right length for the break; a long phone call or a protracted social media session are not.

## Effective note-taking

At this stage in your education, you should be learning to take your own notes. The ability to make clear notes is a vital skill in any job or university. The start of the Diploma course is a good time to take stock of your note-taking talents.

You need to choose whether a hand-written or electronic system is for you. It's also possible to write by hand directly onto a tablet using a stylus. Whichever approach you use, make sure that the notes are permanent and secure. Paper notes should be filed carefully in a sensible order. It is not necessary to carry your entire Diploma notes around in a very large folder. Some students do this and are then unfortunate enough to lose their folder.

With electronic notes it is important to have a regular backup routine to store the files. This storage should be secure and preferably cloud-based. However, if this is not possible, store your files in separate places in several locations. As with notes on paper, this is information that you cannot afford to lose.

Some people scan handwritten notes and store the files electronically. In this case, choose an unambiguous and systematic set of file names.

If you intend to keep notes using a computer or tablet then there are several solutions: programs such as Microsoft OneNote and Google Keep are designed for notes. Once again, choose whatever works for you.

Whether you choose a handwritten or computer-based system, there are several ways to construct lesson notes and the other documents you produce for your schoolwork. When you return to the notes you must be sure that you will still understand them.

### Linear notes

It is tempting to make linear notes during class as they will correspond to the order in which you were taught. However, such notes may be less convenient later should you wish to modify or supplement your notes. One solution is to divide the blank page into sections before you go to class. There is a systematic way to do this, devised by Walter Pauk, a professor at Cornell University. There are many references to Pauk's method on the web. A prepared page of note paper is used (figure 1).

Notes are initially made on the right-hand side of the paper only. You review notes (and therefore the lesson) within 24 hours of the class. At this point you add recall questions to the left-hand column and a summary at the bottom of the page. The questions are meant to help you to recall the material later. For example, if you were being taught about vectors and scalars, one recall question might be "*Distinguish between vector and scalar quantities*". This is likely to be a better recall question than "*What is a vector?*" as it forces you to set up a comparison and therefore more links in your brain.

In the space you can enhance your notes later. You might add questions and solutions that you found difficult at the time, more advanced explanations, or references to helpful textbooks.

### Patterns and other visual notes

Some students find it easier to assimilate notes that are highly visual than plain words. For such students it may be better to base a note-taking system on a pictorial or pattern approach. There are many web references to the technique: a good search phrase is "Pattern notes".

The basic idea is to put the main idea or title in the centre of a blank page and, moving clockwise, write the main points on radial lines. These main points divide into sub-themes and so on. People who respond visually can also use colour in their patterns to differentiate between topics.

Figure 2 shows the beginning of a pattern that was developed from *2.2 Electrical resistance*. Why not copy the pattern, then complete and improve it?

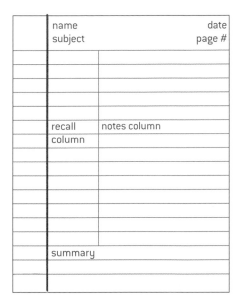

**Figure 1.** The prepared page used in the Cornell note system

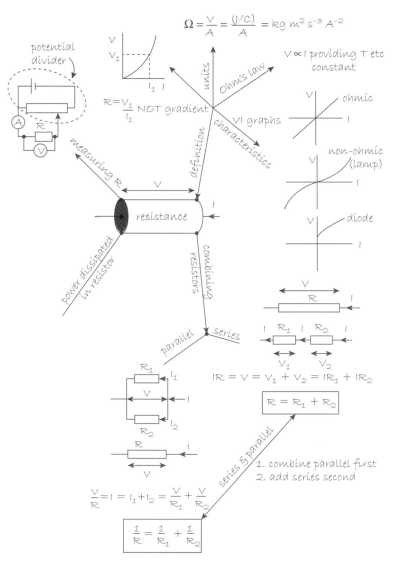

**Figure 2.** A note pattern based on *2.2 Electrical resistance*

*Card notes*

Some students like to make notes on small cards that are roughly 12 cm × 8 cm in size. The advantage here is that such cards are cheap and highly portable. They can be used for quick revision on public transport or in a quiet moment. They can be used either for linear notes, for patterns, or for a combination of both.

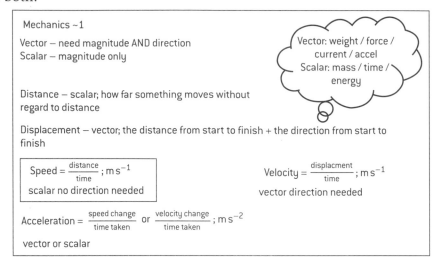

**Figure 3.** A card note based on *1.1 Faster and faster*

## 7.3 Academic honesty

Throughout your Diploma Programme, not just in physics, you will be collecting the words and thoughts of others, from the opinions of classmates through to the writings of distinguished scientists.

It is entirely appropriate that you should quote other people in support of your work. What is not appropriate is passing these words off as your own to gain credit. This is academic dishonesty and, if you are quoting verbatim from others without crediting them, this is plagiarism.

Making full and consistent notes allows you to note down references for later use.

**Internal link**

There is an example of a scientific bibliographic reference in **5.1 Inside the atom.**

## 7.4 Understanding questions

All IB Diploma courses have a set of aims. In physics these concern:

- the factual base, the concepts and the language of physics
- the methods and techniques of physics
- effective communication of physics.

In your Diploma examinations, you will need to be able to:

- **demonstrate** that you have a sound knowledge base of physics, that you understand its conceptual ideas and techniques, and that you can understand and use the language of physics (**Assessment Objective 1**)

- **apply** your knowledge of the subject, its concepts and techniques and **communicate** these (**Assessment Objective 2**)

- **analyse and evaluate** data collected by you or given to you; questions and predictions suggested to you or by you; scientific techniques you may have used or had described to you; scientific explanations provided by you or others (**Assessment Objective 3**)

- **demonstrate practical skills** including research and personal skills to carry through meaningful investigations (**Assessment Objective 4**, which is assessed in the internal assessment, IA).

Notice the increase in difficulty as you move through this list. The term "demonstrate" implies recall of knowledge and ideas, whereas "apply" suggests applying information in some way. The skills of analysis and evaluation are high-order skills that you will spend time developing in class. The questions are designed to tell you which skills are required in a question, and contain a command word. You should note this command and answer accordingly.

Half of all marks in papers 1, 2 and 3 at both SL and HL are devoted to the high-level Assessment Objective 3 (AO3) skills.

## Command terms

The key to answering an examination question well is to read it carefully. Pay special attention to the verb, known as the **command term**, that indicates how you should answer. This command term is normally at the start of the question sentence.

As an example, here is a question taken from chapter 2:

> The graph shows the variation of pd $V$ across resistor A with current $I$ in the resistor.
>
> a) **Calculate** the resistance of A when the current in it is 0.18 A.
>
> b) **Discuss** whether A obeys Ohm's law.
>
> c) Another resistor B has the same resistance as A at a current of 0.18 A. Resistor B obeys Ohm's law. Copy the graph and **sketch** the variation of $V$ with $I$ for resistor B.

The command terms are in bold. These words have a specific meaning in IB Diploma Programme examinations and map onto AO1, AO2 and AO3, telling you the level at which the examiner expects your answer to be pitched. Here are some of the more common terms used together with their meaning. Your teacher can supply you with the complete list.

## Assessment Objective 1 (AO1) – requires recall of facts and concepts of the subject

| Command term | Meaning |
|---|---|
| Define | Give a precise meaning for a word, phrase or idea |
| Draw | Use a labelled, accurate diagram or graph to represent the answer |
| Label | Add labels to a diagram or graph |
| List | Give a sequence of brief answers; no explanation is required |
| State | Give a specific name, value or other brief answer; no explanation is required |
| Write down | Obtain an answer by extracting information; no working need be shown |

These AO1 terms indicate that very little information is required by examiners other than the answer itself. When you see questions like this, do not waste time in providing anything in the way of a detailed explanation.

## Assessment Objective 2 (AO2) – requires application of knowledge and concepts

| Command term | Meaning |
|---|---|
| Annotate | Add brief notes to a diagram or graph |
| Calculate | Obtain an answer showing the relevant stages in the calculation |
| Describe | Give a detailed account |
| Distinguish | Make the difference between items clear |
| Estimate | Obtain an approximate value |
| Identify | Provide an answer from a number of possibilities |
| Outline | Give a brief account or a summary |
| Plot | Mark the position of items on a diagram or graph |

Questions with AO2 terms are often straightforward applications of a piece of knowledge or a single concept. They require an answer that provides the facts but not necessarily a deep analysis of the problem.

## Assessment Objective 3 (AO3) – requires analysis and evaluation of knowledge and concepts

| Command term | Meaning |
|---|---|
| Comment | Make a judgement based on the result of a statement or calculation |
| Compare (and contrast) | Give reasoned similarities (and differences) between two ideas or concepts |
| Deduce | Give all the steps required to reach a conclusion from given facts |
| Demonstrate | Give a reasoned argument using practical or theoretical examples |
| Derive | Produce a new equation from given facts |
| Determine | Obtain a solution giving all the steps involved in the solution |
| Discuss | Give a balanced view that includes a full range of arguments and ideas supported by clear evidence |
| Explain | Give a detailed account that explores all areas of the argument |
| Hence … | Use earlier work in the question to reach a result |
| Justify | Support a conclusion using valid reasons |
| Predict | Reach an expected result giving all steps in the argument |
| Show | Give all the steps involved in a calculation or a reasoned proof of an equation |
| Sketch | Represent a concept or idea using a fully annotated diagram or graph. Note that a higher standard of representation is required than with "Draw". |
| Solve | Arrive at an answer using an algebraic or graphical solution |
| Suggest | Give a solution or hypothesis for a problem |

The AO3 list is longer and shows the increased variety of tasks that you will be asked to attempt. Terms from AO2 and AO3 often appear as pairs, so there is an apparent connection between **calculate/ determine** and **outline/explain**. The terms in the pair have different meanings to do with both the level of explanation required and the complexity of the situation. Compare **(a)** and **(b)** in the following question:

> A boy drags a box across horizontal rough ground at a constant speed.
>
> a)  Outline whether the box is in equilibrium.                    [2 marks]
>
> b)  The boy then drags the box up a hill at the same constant speed. Explain, with reference to the forces acting on the box, why the boy produces a larger average power when moving up the hill.                    [3 marks]

Question **(a)** requires some knowledge of:

* what is meant by equilibrium
* an application of the concept of equilibrium to this question indicating that, because there is no acceleration, the forces are balanced and the box is in equilibrium.

Question **(b)** requires several separate ideas to be combined and analysed. An ideal answer might include:

* the force exerted by the boy must be greater because there is now a gravitational component of weight acting
* friction forces are still present
* power = force × speed
* therefore, as the speed is the same, the power must be greater.

 As you progress through DP Physics, you will need to prepare for your practical *Internal assessment* and for the examinations at the end of the course. Advice for these aspects of your work is found on the book website, at **https://www.oxfordsecondary. com/9780198423591**

# Appendix

## Physical constants

For each paper in the Diploma Programme Physics examination you are provided with a Data Booklet which contains data and equations. Learn how to use this booklet effectively.

- It is worth learning some of the equations by heart. Referring to the data booklet in an examination or test takes time. Some of the equations are part and parcel of understanding the subject.
  Examples of this include *power = force × speed*, the kinematic (*suvat*) equations, Newton's second law ($F = ma$), and so on.

- Learn the most common prefixes such as M, µ, p, n and G. The full list is in *1.3 Work and energy*.

| Quantity and symbol | Approximate value |
|---|---|
| Acceleration due to gravity $g$ | $9.81 \text{ m s}^{-2}$ |
| Speed of light in a vacuum $c$ | $3.00 \times 10^8 \text{ m s}^{-1}$ |
| Charge on the electron $e$ | $1.60 \times 10^{-19} \text{ C}$ |
| Electron rest mass $m_e$ | $9.110 \times 10^{-31} \text{ kg}$ |
| Proton rest mass $m_p$ | $1.673 \times 10^{-27} \text{ kg}$ |
| Neutron rest mass $m_n$ | $1.675 \times 10^{-27} \text{ kg}$ |
| Solar constant (intensity of Sun at orbit of Earth) | $1.36 \times 10^3 \text{ W m}^{-2}$ |

## Unit conversions

1 radian (rad) $\equiv \dfrac{180°}{\pi}$

1 kilowatt-hour $= 3.60 \times 10^6 \text{ J}$

Temperature in kelvin = temperature in degrees Celsius + 273

### A note about radian measure

Although not used in this book, you are likely to meet radians in the course. You will be familiar with sine, cosine and tangent. These are defined in terms of the angles in a right-angle triangle (figure 1 (a)); radians are an alternative angular measure based on the circle (figure 1(b)).

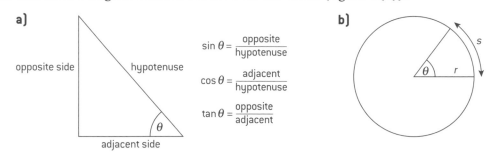

$$\sin\theta = \frac{\text{opposite}}{\text{hypotenuse}}$$

$$\cos\theta = \frac{\text{adjacent}}{\text{hypotenuse}}$$

$$\tan\theta = \frac{\text{opposite}}{\text{adjacent}}$$

**Figure 1. (a)** Trigonometry in a right-angled triangle; **(b)** definition of a radian

$\theta$ in radians is defined as $\dfrac{\text{distance } s \text{ subtended by } \theta}{\text{radius } r \text{ of circle}} = \dfrac{s}{r}$. The radian, abbreviated to rad, is an unusual unit because, like angle degrees, it is a ratio of $\dfrac{\text{length}}{\text{length}}$ and so has no SI units in the usual sense.

When $\theta$ is small, a few degrees, then $\theta$ in rad and $\sin\theta$ are almost the same. When $\theta$ is 90° (a quarter of a circle) then $\theta$ in rad $= \dfrac{s}{r} = \dfrac{\text{circumference of a circle}}{4r} = \dfrac{\pi}{2}$. At this point $\sin 90 = 1$ so the difference is large.

# Equations

## Mechanics (chapter 1)

$v = u + at$

$s = ut + \dfrac{1}{2}at^2$

$v^2 = u^2 + 2as$

$s = \dfrac{(v + u)}{2}t$

$F = ma = m\dfrac{\Delta v}{\Delta t} = \dfrac{\Delta(mv)}{\Delta\ t} = \dfrac{\Delta p}{\Delta t}$

$p = mv$

$\text{impulse} = \Delta p = F \times \Delta t$

$\text{efficiency} = \dfrac{\text{useful energy output}}{\text{total energy input}} = \dfrac{\text{useful power output}}{\text{total power input}}$

$W = Fs\cos\theta$

$\Delta E_k = \dfrac{1}{2}m\left(v^2 - u^2\right)$

$\Delta E_p = mg\Delta h$

$E_k = \dfrac{p^2}{2m}$

$\text{power} = \dfrac{\text{energy transferred}}{\text{time taken}}$

$\text{power} = F \times v$

| | |
|---|---|
| $a$ | acceleration |
| $\Delta E_k$ | change in kinetic energy |
| $\Delta E_p$ | change in gravitational potential energy |
| $F$ | force |
| $g$ | acceleration due to gravity |
| $m$ | mass |
| $p$ | momentum |
| $s$ | distance travelled |
| $t$ | time taken |
| $u$ | initial speed |
| $v$ | final speed |
| $W$ | work done |

## Electricity and magnetism (chapter 2)

$I = \dfrac{\Delta Q}{\Delta t}$

$V = \dfrac{W}{Q}$

$R = \dfrac{V}{I}$

$P = IV = I^2R = \dfrac{V^2}{R}$

series: $R_{total} = R_1 + R_2 + \ldots$

parallel: $\dfrac{1}{R_{total}} = \dfrac{1}{R_1} + \dfrac{1}{R_2} + \ldots$

$R_{total} = \dfrac{R_1 \times R_2}{R_1 + R_2}$

Kirchhoff's laws:

junction: $I_{total} = I_1 + I_2 + \ldots$

loop: $V_{total} = V_1 + V_2 + \ldots$

$\dfrac{V_p}{V_s} = \dfrac{N_p}{N_s} = \dfrac{I_s}{I_p}$

| | |
|---|---|
| $I$ | electric current |
| $n_p$ | number of primary turns |
| $n_s$ | number of secondary turns |
| $Q$ | charge |
| $R$ | resistance |
| $V$ | potential difference *or* electromotive force |
| $V_p$ | primary voltage |
| $V_s$ | secondary voltage |
| $W$ | work done |

## Thermal physics (chapter 3)

$p = \dfrac{F}{A}$

$\rho = \dfrac{m}{V}$

For an ideal gas, $pV = \text{constant}$ when $T$ constant

$\dfrac{p}{T} = \text{constant}$ when $V$ constant

$\dfrac{V}{T} = \text{constant}$ when $p$ constant

| | |
|---|---|
| $A$ | area |
| $p$ | pressure |
| $\rho$ | density |
| $T$ | temperature |
| $V$ | volume |

## Waves (chapter 4)

$T = \dfrac{1}{f}$

$c = f\lambda$

$I \propto A^2$

$I \propto \dfrac{1}{x^2}$

$\dfrac{n_1}{n_2} = \dfrac{v_2}{v_1} = \dfrac{\sin\theta_2}{\sin\theta_1}$

| | |
|---|---|
| $A$ | amplitude (or area depending on the context) |
| $c$ | wave speed |
| $f$ | frequency |
| $I$ | intensity |
| $\lambda$ | wavelength |
| $n_M$ | refractive index in medium M |
| $\theta_M$ | angle between normal and ray in medium M |
| $v_M$ | wave speed in medium M |
| $x$ | distance from source |

## Generating and using energy (chapter 6)

$$\text{maximum power} = \frac{1}{2}\rho A v^3$$

$$\text{maximum power per unit length} = \frac{1}{2}\rho g A^2 v$$

$A$   amplitude (or area depending on the context)

$\rho$   density

$v$   wave speed

## Maths and practical skills

When $l = m \pm n$ then $\Delta l = \Delta m + \Delta n$

When $l = \dfrac{m \times n}{p}$ then $\dfrac{\Delta l}{l} = \dfrac{\Delta m}{m} + \dfrac{\Delta n}{n} + -\dfrac{\Delta p}{p}$

When $l = m^n$ then $\dfrac{\Delta l}{l} = $ magnitude of $n \times \dfrac{\Delta m}{m}$

## Maths skills

This table provides a list of all the skills required for the DP Physics course at both SL and HL, with links to the section and page number of the relevant *Maths skills* box within this book.

| You should be able to: | Section and page number of Maths skills box | Notes |
|---|---|---|
| perform the basic arithmetic functions: addition, subtraction, multiplication and division | 1.1 Faster and faster (page 2) | The assumption is that you can readily perform these operations. Should you need guidance see the section on addition, subtraction, multiplication and division of scalar quantities. |
| carry out calculations involving:<br><br>means (averages)<br><br>decimals<br><br>fractions<br><br>percentages<br><br>ratios<br><br>reciprocals | 3.1 States of matter (pages 74, 78) | A mean is the average of several quantities:<br><br>the mean of $x = 2$, 5 and 8 $= \dfrac{2+5+8}{3} = \dfrac{15}{3} = 5$.<br><br>A decimal is a number expressed to the right of a decimal point and represents a number less than one, e.g. 4.73.<br><br>A fraction is a number expressed using a numerator and a denominator, e.g. $\dfrac{1}{5}$.<br><br>A percentage is a number expressed as a fraction of one hundred, e.g. 3 in 5 is $\dfrac{3}{5} \times 100 = 60\%$.<br><br>A ratio is the number of times one number is contained in another, e.g. $x : y = 1 : 8$ means 1 $x$ for every 8 $y$.<br><br>A reciprocal is a number divided into one, e.g. the reciprocal of 4 is $\dfrac{1}{4} = 0.25$. |
| represent arithmetic mean using x-bar notation | 3.1 States of matter (page 74) | When $x = 2$, 5 and 8, $\bar{x} = 5$ |
| use radian measure | Appendix (page 169) | |
| use trigonometric functions | 1.2 Pushes and pulls (pages 18, 20), 4.2 Physics of light (page 108), Appendix (page 169) | Use of sine, cosine, tangent and $\sin^{-1}$, $\cos^{-1}$ and $\tan^{-1}$ to find angles and lengths<br><br>Covered in the *Maths skills* sections |
| use standard form (for example, $3.6 \times 10^6$) | 1.1 States of matter (page 4), 1.3 Work and energy (page 21) | Standard notation is in form $X.YZ \times 10^p$ where $X$, $Y$, $Z$ and $p$ are integers.<br><br>Engineering notation only allows $p$ to be divisible by 3, so it can only be 3, 6, 9, 12, etc. |

| | | |
|---|---|---|
| solve simple algebraic equations | 3.3 Changing temperature and state (page 88) | Covered in the *Maths skills* section |
| solve linear simultaneous equations | 3.3 Changing temperature and state (page 88) | Covered in the *Maths skills* section |
| use direct and inverse proportion | 3.2 Gas laws (page 82) | When $x \propto y$ then a graph of $y$ against $x$ is a straight line through the origin – this is direct proportion.<br><br>When $x \propto \dfrac{1}{y}$ then a graph of $\dfrac{1}{y}$ against $x$ is a straight line<br><br>through the origin – this is inverse proportion.<br><br>Covered in the *Maths skills* section |
| plot graphs (with suitable scales and axes) including two variables that show linear and non-linear relationships | 1.1 Faster and faster (page 12) | Covered in the *Maths skills* section |
| interpret graphs, including the significance of gradients, changes in gradients, intercepts and areas | 3.2 Gas laws (page 82) | Covered in the *Maths skills* section |
| draw lines (either curves or linear) of best fit on a scatter plot graph | 1.1 Faster and faster (page 12), 5.3 Half-life (page 136) | Covered in the *Maths skills* sections |
| on a best-fit linear graph, construct linear lines of maximum and minimum gradients with relative accuracy (by eye) considering all uncertainty bars | 5.3 Half-life (page 136) | <br><br>Draw maximum and minimum gradient lines that touch all the error bars – the values of these gradients indicate the range of error in the gradient. |
| interpret data presented in various forms (for example, bar charts, histograms and pie charts) | There is a pie chart featured in 5.2 Radioactive decay (page 129) | *Bar chart:* A graph that presents data with rectangular bars whose heights are proportional to the values represented. Good for categoric variables.<br><br>*Histogram:* A graph that is made of rectangles with an area proportional to the frequency of the variable and a width equal to the class interval. Good for ranges.<br><br>*Pie chart:* A circle divided into sectors that represent the fraction of the whole chart and population. |
| express uncertainties to one or two significant figures, with justification | 1.1 States of matter (page 4), 2.2 Electrical resistance (page 48) | *In questions:* express answers to the **same** degree of precision as the data in a question. Round your answer at the last stage in a calculation.<br><br>*In practical work:* express answers to the same number of significant figures as indicated by error determinations.<br><br>Covered in the *Maths skills* sections |

# Index

Key terms are in **bold**.